Quantum Chemical and Molecular Dynamics Simulations for Beginners

This book provides an in-depth exploration of the dynamic field of computational chemistry. Spanning seven meticulously crafted chapters, it begins by laying a solid foundation in quantum chemistry and electronic structure theory, introducing essential tools and software that underpin computational chemistry calculations. The journey continues with a comprehensive examination of transition state analysis for interpreting reaction mechanisms, followed by an insightful discussion on modern solvation studies using QM/MM techniques and the practical implementation of hybrid models. This book concludes with an engaging dive into molecular dynamics simulations, enriched by real-world case studies utilizing NAMD and VMD to illustrate the complex behavior of molecular systems.

Key features:

- Focuses on quantum chemical calculations designed for beginners
- Reviews the use of atomistic molecular dynamics simulations for various systems
- Presents detailed analysis of calculations and result analysis on real systems
- Includes usage of widely accepted software packages such as Gaussian and NAMD
- Explores basic parameters associated with designing a chemical system

This book is aimed at researchers and senior undergraduate students in chemical engineering, chemistry, and materials science.

Quantum Chemical and Molecular Dynamics Simulations for Beginners

Dhirendra Kumar Mishra, Nabendu Paul,
Nipu Kumar Das, Nikhil Kumar,
and Tamal Banerjee

CRC Press
Taylor & Francis Group
Boca Raton London New York

CRC Press is an imprint of the
Taylor & Francis Group, an **informa** business

Designed cover image: Getty Images

First edition published 2026
by CRC Press
2385 NW Executive Center Drive, Suite 320, Boca Raton FL 33431

and by CRC Press
4 Park Square, Milton Park, Abingdon, Oxon, OX14 4RN

CRC Press is an imprint of Taylor & Francis Group, LLC

ISBN: 9781032619286 (hbk)
ISBN: 9781032694757 (pbk)
ISBN: 9781032694733 (ebk)

DOI: 10.1201/9781032694733

Typeset in Times
by codeMantra

Contents

Acknowledgments

The authors are grateful to GAUSSIAN, NAMD, VMD, and Multiwfn to allow the use of their logo and images for this book. NAMD was developed by the Theoretical and Computational Biophysics Group in the Beckman Institute for Advanced Science and Technology at the University of Illinois at Urbana-Champaign. The authors also would like to thank the Ministry of Education, Government of India for allowing us to adapt/reuse a part of my course from NPTEL repository namely "Molecular Simulations in Chemical Engineering". The following citations refers to the software's and their citations which has been used thoroughly in the book:

1. GaussView, Version 6, Dennington, Roy, Keith, Todd A., Millam, John M. Semichem Inc., Shawnee Mission, KS, 2016.
2. Gaussian 09, Revision D.01, Frisch, M. J., Trucks, G. W., Schlegel, H. B., Scuseria, G. E., Robb, M. A., Cheeseman, J. R., Scalmani, G., Barone, V., Petersson, G. A., Nakatsuji, H., Li, X., Caricato, M., Marenich, A. V., Bloino, J., Janesko, B. G., Gomperts, R., Mennucci, B., Hratchian, H. P., Ortiz, J. V., Izmaylov, A. F., Sonnenberg, J. L., Williams-Young, D., Ding, F., Lipparini, F., Egidi, F., Goings, J., Peng, B., Petrone, A., Henderson, T., Ranasinghe, D., Zakrzewski, V. G., Gao, J., Rega, N., Zheng, G., Liang, W., Hada, M., Ehara, M., Toyota, K., Fukuda, R., Hasegawa, J., Ishida, M., Nakajima, T., Honda, Y., Kitao, O., Nakai, H., Vreven, T., Throssell, K., Montgomery, J. A., Jr., Peralta, J. E., Ogliaro, F., Bearpark, M. J., Heyd, J. J., Brothers, E. N., Kudin, K. N., Staroverov, V. N., Keith, T. A., Kobayashi, R., Normand, J., Raghavachari, K., Rendell, A. P., Burant, J. C., Iyengar, S. S., Tomasi, J., Cossi, M., Millam, J. M., Klene, M., Adamo, C., Cammi, R., Ochterski, J. W., Martin, R. L., Morokuma, K., Farkas, O., Foresman, J. B., Fox, D. J. Gaussian, Inc., Wallingford CT, 2016.
3. Lu, T., & Chen, F, Multiwfn: A multifunctional wavefunction analyzer. *Journal of Computational Chemistry*, 2012. **33**(5): p. 580–592. https://doi.org/10.1002/jcc.22885
4. Phillips, J. C., Hardy, D. J., Maia, J. D. C., Stone, J. E., Ribeiro, J. V., Bernardi, R. C., Buch, R., Fiorin, G., Hénin, J., Jiang, W., McGreevy, R., Melo, M. C. R., Radak, B. K., Skeel, R. D., Singharoy, A., Wang, Y., Roux, B., Aksimentiev, A., Luthey-Schulten, Z., Kalé, L. V., Schulten, K., Chipot, C., & Tajkhorshid, E, Scalable molecular dynamics on CPU and GPU architectures with NAMD. *The Journal of Chemical Physics*, 2020. **153**(4): p. 044130. https://doi.org/10.1063/5.0014475
5. Humphrey, W., Dalke, A. and Schulten, K., VMD - visual molecular dynamics. *Journal of Molecular Graphics and Modelling*, 1996. **14**: p. 33–38.

About the Authors

Dhirendra Kumar Mishra earned a Doctorate in Chemical Engineering from the Indian Institute of Technology, Guwahati, Assam, India. During his doctoral studies, he specifically concentrated on the integration of ionic liquids and eutectic solvents in the field of hydrogen storage as a catalytic solvent. His study involved utilizing experimental and computational chemistry simulations to comprehend and improve the effectiveness of ionic liquids and eutectic mixtures for practical usage.

His research interests span multiple areas, including computational chemistry for energy-related applications, green and sustainable solvents, hydrogen storage, and nanofluids for thermal applications. He is currently a postdoctoral fellow at the Joint BioEnergy Institute, affiliated with Sandia National Laboratories, where his research focuses on improving the process of catalytic lignin depolymerization using solvents and heterogeneous catalysts through experimental and computational approaches.

He has published more than ten research papers in journals with high impact factors, including *Chemical Reviews, ACS Sustainable Chemistry and Engineering, Journal of Physical Chemistry A*, *International Journal of Hydrogen Energy*, and the *Journal of Energy Storage.*

Nabendu Paul earned his Doctoral degree from the Chemical Engineering Department of the Indian Institute of Technology, Guwahati, in 2023. Subsequently, he joined the Maulana Azad National Institute of Technology (MANIT), Bhopal, and has been serving as an Assistant Professor in the Department of Chemical Engineering since January 2024. He obtained his M. Tech. and B. Tech. degrees in Chemical Engineering from the National Institute of Technology, Agartala, in 2015 and 2013, respectively.

His research encompasses a wide range of interdisciplinary domains, including ionic liquids and deep eutectic solvents (DESs), extraction and recovery of pharmaceuticals and biomolecules from aqueous environments, molecular dynamics and quantum chemical calculations, and the study of non-covalent interactions. His work further extends to ionic liquid-based systems for environmental and biological applications, such as drug delivery, direct air capture (DAC) of CO_2, and the development of eutectogels, ionogels, and DES-based electrolytes for supercapacitors. Recently, his group has initiated studies on the antibacterial potential of DESs in wound healing membranes.

He has published over eight articles in reputed peer-reviewed journals, including *Small, ACS Sustainable Chemistry and Engineering, Journal of Physical Chemistry B,* and the *Journal of Molecular Liquids*. He has also authored a book that discusses the theoretical correlations and molecular dynamics aspects of deep eutectic solvents in environmental applications. Additionally, he has published several book chapters and conference papers. His work integrates both experimental and computational approaches and contributes to advancing sustainable and functional material systems in chemical and environmental engineering.

Nipu Kumar Das received his Doctoral degree in 2025 from the Indian Institute of Technology, Guwahati, Assam, India, in Chemical Engineering. During his Ph.D. studies, Dr. Das primarily focused on deep eutectic solvent-based thermal fluids and solubility studies of various natural deep eutectic solvents. His studies involved both experimental and computational chemistry to understand the mechanism of solvent interactions with nanoparticles.

His other research interests span the areas of green nanoparticles, energy-related applications, COSMO-SAC modeling, corrosion inhibitors, and sustainable solvents. He has published more than 13 research papers in journals with high impact factors, including the *Journal of Molecular Liquids*, *ACS Sustainable Chemistry and Engineering*, *AIChE Journal*, *Industrial Crops and Products*, and the *Journal of Energy Storage.*

Nikhil Kumar is a postdoctoral researcher at Sandia National Laboratories who is currently working in the Predictive Deconstruction group at the Department of Energy-funded Joint BioEnergy Institute (JBEI). Dr. Kumar received his Ph.D. in 2023 from the Department of Chemical Engineering at the Indian Institute of Technology Guwahati, India.

Dr. Kumar's research is primarily focused on the development of novel green solvents for aromatic compound extraction and understanding the liquid structure of deep eutectic solvents using molecular dynamics, quantum chemistry, and COSMO-RS model simulations. Dr. Kumar's research interests also include leveraging both machine learning and first-principles physics-based methods to make improved predictions for solvents for enhanced biomass conversion, utilization of CO_2 for the circular economy, and thermodynamic property predictions using various machine learning and artificial intelligence models.

Dr. Kumar has co-authored a book, published more than ten research articles in leading academic journals, and contributed to over ten conferences.

Tamal Banerjee is currently a Professor at the Department of Chemical Engineering, Indian Institute of Technology, Guwahati. He served as Supplemental Faculty in the Department of Chemical and Biomolecular Engineering at the University of Delaware, USA, from 2011–2012. He also held the position of University Associate at Curtin University, Perth, Australia, from 2020–2024. Throughout his career, Dr. Banerjee has authored three books (published by CRC and Elsevier) and around 200 academic publications. His research group has contributed around 900 data points related to thermodynamic properties of ionic liquid (IL) and deep eutectic solvent (DES) mixtures, now available in the NIST database. He has supervised 20 doctoral students, most of whom focus on applications of green solvents in energy and environmental domains.

Dr. Banerjee received the INDO-US Fellowship in 2011 in Engineering Sciences. He is currently a Fellow of the Royal Society of Chemistry, the Indian Institute of Chemical Engineers, and the Indian Chemical Society. He serves on the editorial boards of *Fluid Phase Equilibria* and the *Journal of Solution Chemistry*, and is the Associate Editor of *Thermal Advances.*

His research uses both quantum chemistry and molecular dynamics to predict thermodynamic and transport properties of ILs and DESs. His expertise includes the COnductor like Screening MOdel (COSMO) based methods and molecular modelling techniques, encompassing both classical molecular dynamics and reactive force field simulations (ReaxFF). His application areas span renewable energy sources (such as alcohols) and non-renewable energy sources (such as coal and chemical hydrides).

1 Introduction to Gaussian and GaussView Package, Geometry Optimization, and Frequency Analysis

1.1 INTRODUCTION

1.1.1 Electronic Structure Theory

Computational chemistry is a comprehensive phrase that denotes the application of computer methods to the direct exploration of chemical issues. Consequently, it includes the entire range of computational techniques employed in applied chemistry, regardless of their origins in physics, mathematics, informatics, or other core scientific fields (e.g., quantum mechanics, statistical mechanics).

Computational chemistry techniques can be employed to compare experimental results with predicted molecular properties, clarify ambiguous experimental data, and simulate transient, unstable transition states and intermediates that cannot be directly observed.

The Schrodinger equation (Equation 1.1) is the primary reference, making quantum chemistry a more accurate term for methods that are partially or wholly based on the fundamental principles of quantum mechanics.

$$H\psi = E\psi \tag{1.1}$$

The field of quantum chemistry can be essentially divided into two main sections:

- Techniques whereby the nucleus of a molecule is considered a stationary entity encircled by moving electrons. This assumption is reasonable given that electrons move far faster than nuclei. These approaches are also referred to as "electronic structure methods" since their emphasis is on electrons.
- Methods for approximating molecular behavior over time. Advanced methods in the study of chemical processes address the complicated behaviors of electrons and nuclei using the foundations of quantum mechanics (molecular dynamics), classical Newtonian mechanics (quantum dynamics), or a combination of both (semiclassical dynamics).

One can arrange methods for electronic structure into several categories. Among others, a group of people depends just on the ideas and values of fundamental constants

DOI: 10.1201/9781032694733-1

in quantum physics, such the speed of light, the masses, and charges of electrons and nuclei. Several approaches combine parameters obtained from experimental results into variants of the same method in order to simplify the computations. Some people build molecular systems combining components from several approaches. The methods show variations in the trade-offs made between computing complexity and accuracy.

In cases when parameter sets are precisely described, for instance, semi-empirical computations offer reasonably accurate quantitative predictions of energies and structures together with realistic qualitative descriptions of molecular systems, all at a similar cost. Not confined to particular materials, system types, or environmental circumstances, DFT calculations offer exact predictions of high quality over a wide spectrum of systems.

Many times, software for electronic structure offers molecular mechanics approaches as an additional set of tools. Driven by conventional ideas, MM techniques are quite effective and applicable to systems with millions of atoms. Methods in molecular mechanics show the potential energy of molecules depending on interactions between atomic pairs by using force fields composed of parameter sets and particular functional forms. These methods are mostly applied in the field of electronic structure computations to replicate very big molecules by aggregating molecular mechanics with quantum mechanics.

Model chemistry can be defined most simply as the mix of a theoretical approach with a basis set. Forecasting distinct processes or molecules using a consistent model method is crucial for efficient comparison of their features. These descriptions have as their goal simplicity. With each method showing different degree of accuracy, several theoretical approaches are connected to special approximations of the Schrodinger equation, cost of computation, and required supplies. Levels of theory are different theoretical approaches where more levels suggest more clarity.

The Gaussian 09/16 software package includes a variety of theoretical methods applicable to chemical systems and molecular states with varying levels of accuracy. The following are notable families of procedures available in Gaussian:

Semi-empirical methods are cost-effective qualitative approaches. They are beneficial for the preliminary modeling of big molecules, which are impractical to model using more exact methods, and as constituents in ONIOM MO:MO and MO:MM calculations.

Illustrations:
ZINDO: This technique was parameterized to replicate electronic spectra. It is primarily utilized for calculating UV transitions.
AM1: Austin Model 1 is a technique predominantly employed to represent organic compounds.

The Hartree–Fock method is the most accurate and economical ab initio methodology. It was the favored preliminary approach for an extended duration. The Hartree–Fock method is constrained in its ability to account for electron correlation effects, which denote the energy contributions arising from electron interactions. Molecular orbitals can be computed using two methods in HF: UHF (unrestricted) or RHF (restricted). UHF allocates a distinct orbital for each electron, regardless of pairing, applicable to ions, excited states, radicals, and similar entities. RHF employs the same orbital spatial function for electrons within the same pair, which is advantageous for species with paired electrons, hence eliminating spin contamination. The primary limitation of the HF approach is the omission of electron correlation. The subsequent models commence with a Hartree–Fock calculation and subsequently adjust for electron repulsion.

Moller–Plesset perturbation theory is based on the subfield of many-body perturbation theory within mathematical physics. These techniques enhance Hartree–Fock theory with corrections, yielding markedly improved precision for systems where electron correlation effects are substantial, albeit at a proportionate increase in expense. In reality, only MP2 and MP4 methods are utilized, as alternative methods are either excessively computationally intensive or do not substantially enhance the results relative to a lower complexity level.

Configuration interaction calculations are predominantly employed for excited states. CI can achieve high accuracy; nevertheless, they are also highly demanding on CPU resources.

Density functional theory (DFT): These methodologies include many electron correlation effects with no additional expense compared to Hartree–Fock theory, leading to its widespread replacement. A multitude of DFT methodologies exists. The differentiation between pure and mixed approaches is substantial. A pure technique consists of two functionals, specifically exchange and correlation, which delineate two separate components in the energy expression. Hybrid functionals are formulated by including a portion of the Hartree–Fock exchange term. Dispersion effects and long-range interactions (electron–electron interactions that transpire at significant distances from the nucleus during events such as electron excitations to elevated orbitals) have lately been incorporated into advancements in DFT. DFT differs from HF-based approaches in that it uses electron density to compute energy rather than a wave function.

Example:
B3LYP: This is the preeminent DFT model. This method is termed hybrid because it incorporates corrections for both gradient and exchange correlations.
PW91: This is a frequently utilized exchange-correlation functional. The essential element is the estimation of the exchange and correlation energy among electrons in a many-electron system.

Coupled cluster methods integrate many levels of electron excitations, entailing the interchange of numerous real and virtual orbitals inside the Hartree–Fock wavefunction. This yields remarkably great precision, however, at a considerable cost.

Compound models approximate highly accurate methods for forecasting thermochemistry via a sequence of more economical calculations. The complete basis set (CBS) and Gaussian-n (Gn) technique families exemplify such models.

1.1.2 Basis Sets

1.1.2.1 What Is a Basis Set and Why Is Its Selection Important?

A basis set comprises wave functions that characterize the configuration of atomic orbitals (AOs). The molecular orbitals (MOs) are calculated using the chosen theoretical model by linear combination of atomic orbitals (LCAO). Not all theoretical models necessitate the selection of a basis set by the user. For instance, PM_n ($n=3, 6$) models utilize an internal basis set, but ab initio or DFT necessitate a specified basis set. The accuracy of your calculation is closely correlated with the employed basis set. The decision involves a trade-off between result accuracy and CPU time. Slater-type orbitals (STOs) and Gaussian-type orbitals (GTOs) are employed to characterize AOs. STOs more accurately depict the morphology of AOs compared to GTOs; yet, GTOs possess an unparalleled advantage due to their computational simplicity. Indeed, it is more expedient to calculate multiple GTOs and amalgamate them to characterize an orbital than to compute a single STO. Combinations of GTOs are frequently employed to characterize STOs, which then define AOs.

Note: For detailed theory about Quantum Chemistry and its subsections visit Gaussian Website: https://gaussian.com/

1.1.3 Getting Started with Gaussian and GaussView

1.1.3.1 Introduction to Gaussian and GaussView Package

Gaussian 09 and GaussView 5.0 (commercially available versions) represent a powerful combination of software that is widely used in both academic and industrial fields to apply quantum mechanical methods in the study of molecular structures, properties, and reactions. Here is a concise overview of both:

1.1.3.1.1 Gaussian 09 Software Suite

The software suite Gaussian 09, designed for computational analysis, was created by Gaussian, Inc. Employing fundamental quantum techniques, it offers a comprehensive range of capabilities for exploring molecular systems. Key features of Gaussian 09 consist of

Calculations of Electronic Structure: Gaussian 09 excels in executing a wide variety of electronic structure calculations, including Hartree–Fock (HF), DFT, and molecular orbital (MO) calculations, among others.

Properties of Molecules: A comprehensive array of molecular characteristics can be determined, encompassing thermodynamic attributes, energy levels, structural configurations, vibrational frequencies, and electronic spectra.

Evaluation of Reaction Pathways: Gaussian 09 provides the tools to explore reaction pathways, transition states, and mechanisms through techniques such as intrinsic reaction coordinate (IRC) calculations and transition state optimization.

Solvation Models: To incorporate the influence of solvents in computations, solvation models are utilized.

Versatile Input Choices: Gaussian 09 offers the ability to customize calculation parameters and accommodates input in multiple formats.

Parallel Processing: This capability allows users to accelerate calculations by allocating them across multiple processing units or nodes. GaussView 5.0 Software Suite **GaussView 5.0** functions as the graphical user interface (GUI) for Gaussian 09. It provides a straightforward and accessible platform for setting up, running, and interpreting Gaussian 09 calculations.

The main features of the GaussView package include:

Input Preparation: GaussView allows users to visually create molecular structures, set calculation parameters, and produce input files for use in Gaussian computations.

Visualization: The software offers tools that allow users to examine properties calculated using Gaussian methods, including molecular structures, orbitals, and electron densities.

Job Monitoring: GaussView provides users with the ability to view intermediate results and track the progress of calculations in real time.

Analysis Tools: It offers insights into molecular characteristics, vibrational frequencies, electronic spectra, and various instruments for examining Gaussian output files.

Graphical Output: GaussView generates visual representations such as electron density diagrams, molecular orbitals, and electrostatic potentials, along with various other graphical depictions of computed properties.

Integration: The application effortlessly connects with Gaussian 09, allowing users to start Gaussian calculations directly from the GaussView interface and retrieve the resulting data without leaving the application.

Note: It is important to recognize that Gaussian 9 and GaussView 5.0 are proprietary software applications, and typically, users must obtain licenses to access their features. These software packages are developed and distributed by Gaussian, Inc., a computational software company based in the United States.

Gaussian, Inc. offers a range of licensing options and pricing structures tailored for academic, government, and commercial users, considering factors such as the number of users, type of institution, and specific features desired. For details on pricing, licensing, and purchasing options for Gaussian and GaussView, please reach out to Gaussian, Inc. directly or check their official website.

1.1.3.1.2 Setting up Calculation

- Building the molecule in GaussView visualization program
- Setting up the Gaussian job using package menu and dialogs
- Submitting the job

When you initially begin a GaussView session by opening the program, you will see two windows similar to those in the illustration following. The top portion of the GaussView main window contains the program's toolbars and the icons on the left side of the top row control atom/fragment selection for the building process (Figure 1.1).

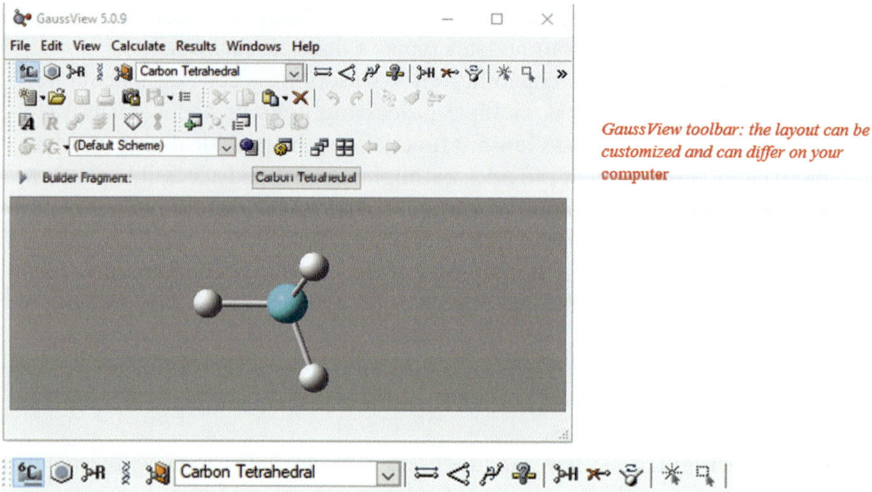

FIGURE 1.1 GaussView program toolbar displaying various tool icons for molecular modeling, including atom selection and bond settings.

We will begin the build process by selecting the hybridization, and selection will show up in the View window:

Select atom from the pop-up menu (tetrahedral carbon as in the figure). The symbols alongside correspond to bonds, dihedrals, and angles (Figure 1.2).

- We can also add a desired atom. Click on the Select Atom icon in the upper left corner of the toolbar. A window containing the periodic table will appear; use it to select the desired element and the bare atom fragment form available at the bottom.

1.1.3.2 Preparing and Submitting a Gaussian Job in GaussView

- Select Gaussian Calculation Setup from the Calculate menu. This will open the corresponding dialog (the image below).
- Select the Title panel and enter a brief description for the job into the Job Title area. Example: Ethanol Optimization (Figure 1.3).
- Select the Method panel. This panel is used to specify the model chemistry or in other words, the theoretical method and basis set used to solve the chemistry of this system. For details associated with each tab, one can visit the Gaussian Website: https://gaussian.com/techsupport/
- The second row of popup menus specifies the desired basis set for the calculation; the basis set name along with any additional diffuse functions (second popup) and/or polarization functions: on heavy atoms (third popup) and on hydrogen atoms (fourth popup).
- After verifying that the charge and spin multiplicity selected by GaussView are correct, we are ready to run the job. Begin the process by clicking on the Submit button. GaussView will first ask whether you want to save

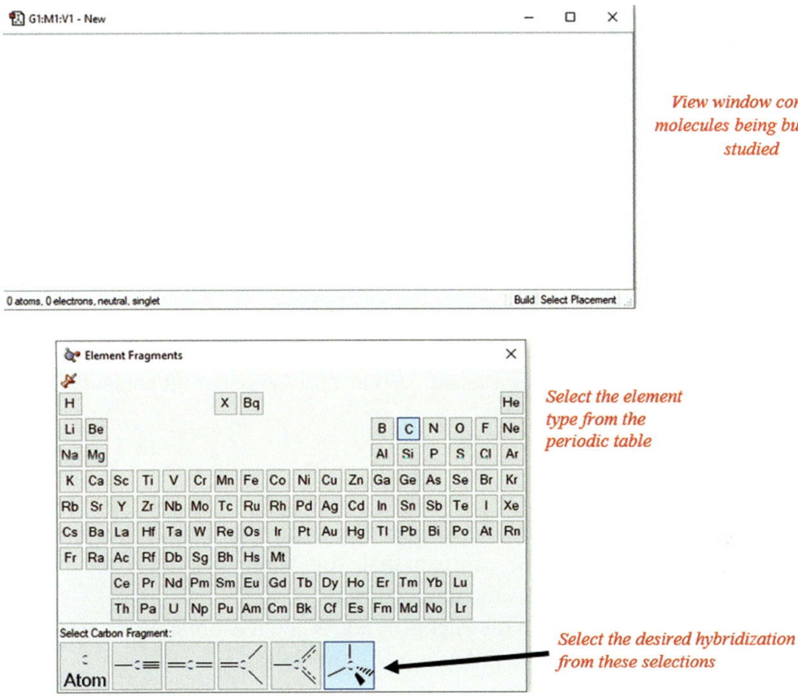

FIGURE 1.2 The GaussView molecule editor toolbar with icons for atom, bond, angle, and dihedral selection.

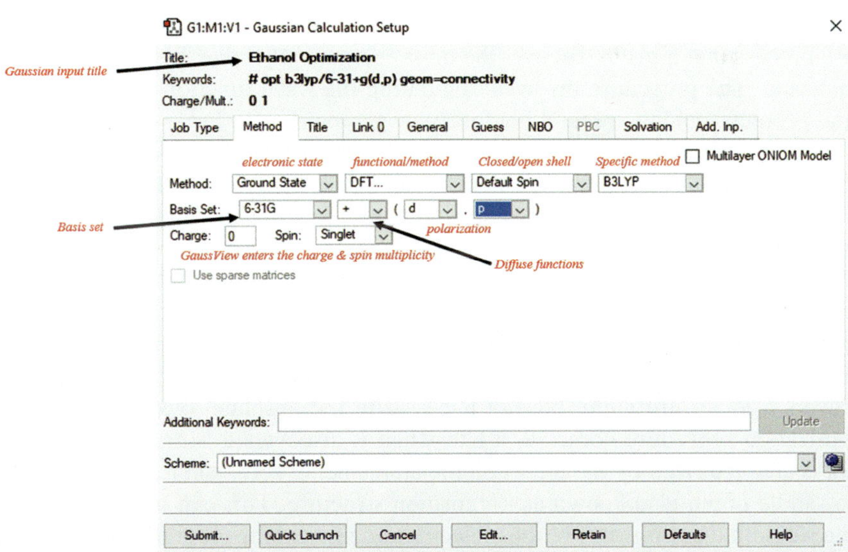

FIGURE 1.3 Interface of the Gaussian Calculation Setup dialog box in GaussView, showing options for setting up electronic structure calculations.

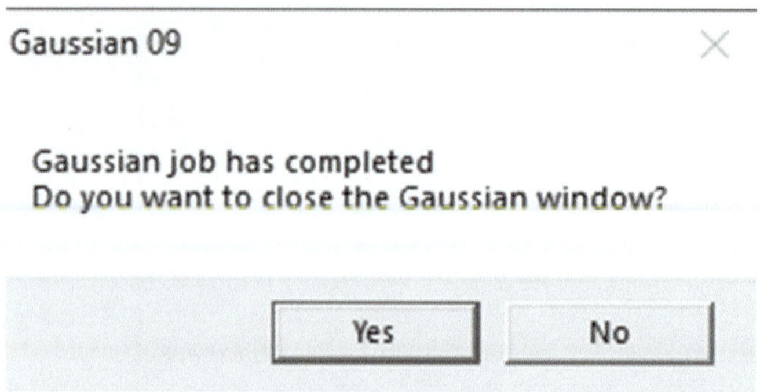

FIGURE 1.4 Dialog box indicating the completion of a Gaussian 09 computation, with options to view or save results.

the Gaussian input file it has generated. Answer Save to this prompt and then select a name and location for the file. Once the file has been saved. GaussView will ask you to confirm that you want to submit the file to Gaussian; select Ok to begin the calculation.

- The Gaussian job will run in the background. When it is finished, GaussView will display a dialog like the following (Figure 1.4):

1.2 THEORY OF GEOMETRY OPTIMIZATION AND FREQUENCY ANALYSIS

1.2.1 Geometry Optimization

It is a process that aims to discover the most energetically favorable configuration of a molecule. The procedure involves calculating the wave function and energy at an initial geometry and then searching for a new geometry with lower energy. This process is repeated until the geometry with the lowest energy is discovered. The procedure determines the force on each atom by analyzing the gradient (first derivative) of the energy in relation to atomic positions. Advanced algorithms are then employed at each stage to choose a new geometry, with the goal of quickly reaching the geometry with the lowest energy. In the final, minimum energy geometry, the force on each atom is balanced and at equilibrium. It is crucial to acknowledge that this procedure may not identify the global minimum, which refers to the geometry with the lowest energy. Through its inherent characteristics, a sequential exploration for a minimum uncovers a local minimum, but not necessarily the absolute lowest. Actually, the optimization procedure comes to a halt when it discovers a stationary point. This point is characterized by the forces on the atoms being zero, and it could potentially be a saddle point, also known as a transition structure. This can happen especially when we limit the symmetry of the molecule and prevent the program from exploring all possible molecular configurations. Beginning a procedure for a geometry optimization calculation with a modest basis set and a less advanced method is a prudent approach before transitioning to the preferred basis set and method for a specific problem. After selecting the geometry using a simpler and less accurate approach, you can proceed with the final geometry optimization. Utilizing the second derivatives of the energy in relation to atomic coordinates, which can be obtained from the optimization process or a frequency run, can be advantageous in enhancing the geometry optimization [1].

1.2.2 Frequency Analysis

Frequency analysis is a crucial computational chemistry technique used to study the vibrational properties of molecules. After geometry optimization, frequency calculations determine whether a molecular structure corresponds to a true energy minimum or a transition state by analyzing the curvature of the potential energy surface. This method is based on the harmonic oscillator approximation, where molecular vibrations are modeled as small displacements around the optimized geometry. The second derivative of the energy with respect to atomic positions forms the Hessian matrix, which provides vibrational frequencies and normal modes of motion. These normal modes describe the collective movement of atoms in stretching, bending, and **torsional motions**.

A molecule with N atoms has $3N$ degrees of freedom, but only $3N-6$ (for nonlinear molecules) or $3N-5$ (for linear molecules) correspond to vibrational motions, as three translational and rotational degrees are removed. The calculated vibrational frequencies are expressed in wavenumbers (cm^{-1}) and can be used to predict infrared (IR) and Raman spectra based on dipole moment and polarizability changes, respectively. These spectral predictions help identify functional groups and molecular fingerprints in experimental spectroscopy.

One of the most critical aspects of frequency analysis is the identification of imaginary frequencies. If all vibrational frequencies are positive (real values), the structure is a stable minimum on the potential energy surface. However, if one or more frequencies are negative (imaginary), the structure corresponds to a saddle point. A single imaginary frequency indicates a transition state, a key feature in reaction pathway analysis, while multiple imaginary frequencies suggest a higher-order saddle point or an unstable configuration.

In addition to vibrational analysis, frequency calculations provide thermodynamic properties such as zero-point energy (ZPE), enthalpy, entropy, and Gibbs free energy. These values are derived using statistical mechanics and are crucial for computing reaction energies, equilibrium constants, and activation barriers. Frequency calculations also include thermal corrections, ensuring accurate predictions of thermodynamic parameters at different temperatures. However, due to limitations in the harmonic approximation, computed frequencies are often scaled using empirical factors to match experimental data more closely.

The accuracy of frequency calculations depends on the choice of quantum chemical method and basis set. DFT methods such as B3LYP, PBE, or hybrid functionals, as well as post-Hartree–Fock methods such as MP2, are commonly used to balance accuracy and computational cost. Before performing frequency analysis, the molecular geometry must be well optimized, as small deviations from the equilibrium structure can lead to incorrect vibrational modes.

What information do you get out of this calculation?

- Atomic coordinates of optimized molecule
- Optimized parameters: atomic distances, dihedrals, and angles
- Orbital analysis (HOMO/LUMO)
- Mulliken atom charges
- Dipole moments

1.2.3 Example: Geometry Optimization of Ethanol Molecule

Step 1: Preliminary steps would be to draw the ethanol structure using GaussView as detailed in previous sections shown in Figure 1.2.

Step 2: In the next step, press right mouse button and a tab will appear. Select calculate followed by selecting Gaussian calculation setup and press left mouse button or press "Ctrl+G" (Figure 1.5 and Figure 1.7) .

Step 3: In the next step, select Job type among available options (Figure 1.7).

Step 4: Select Optimization as the job type (Figure 1.8).

Step 5: Select method and basis sets to be used for optimization of geometry (Figure 1.9).

Step 6: Name the job to be performed in the Title tab (Figure 1.10).

Step 7: Using the Options tab, the user can be able to edit the following mentioned options related to checkpoint file location, memory limit to be used, and processors.

Finally, after the geometry optimization is over, we can open ethanol.out file, which is also named as .log file, consisting of all information related the optimized

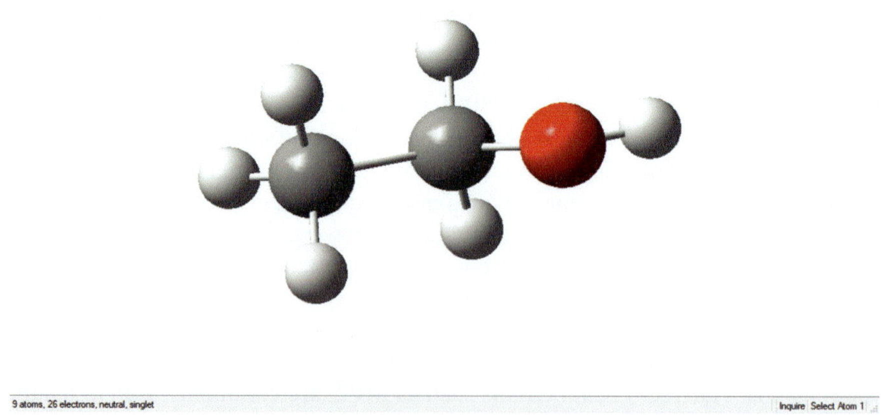

FIGURE 1.5 Visual representation of the ethanol molecule structure in GaussView, displaying atomic connectivity.

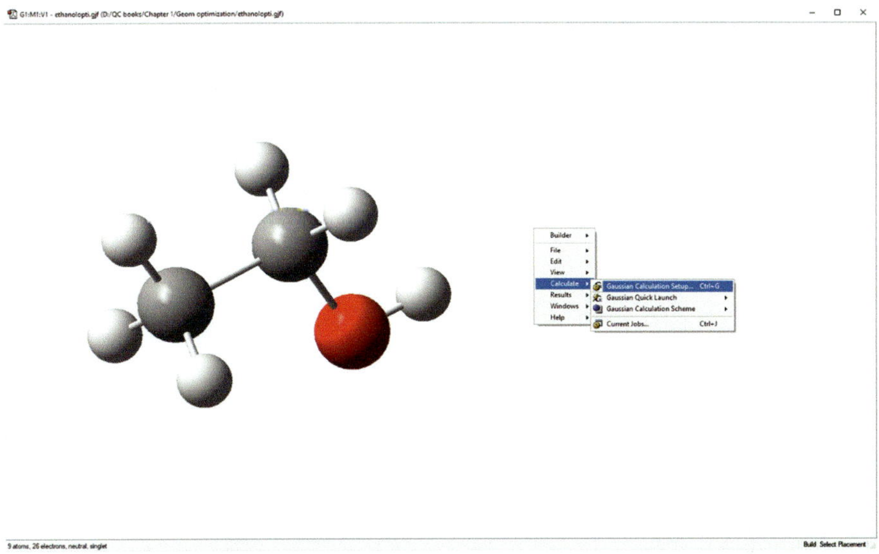

FIGURE 1.6 Detailed view of the GaussView graphical user interface during the setup of a geometry optimization calculation for an ethanol molecule.

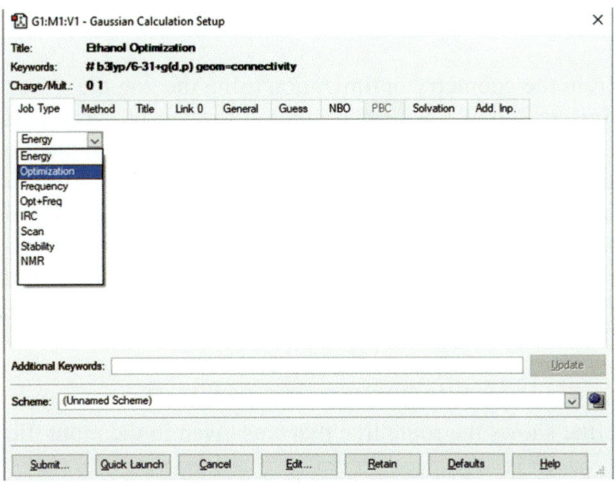

FIGURE 1.7.1 Close-up of the job type selection panel in GaussView, focusing on optimization settings.

geometry detailing about the energy of system, angles, dihedrals, and bond distances between atoms (Figures 1.11 and 1.12).

Step 8: Press right click on the mouse, Go to "Results" and then select "Summary" (Figure 1.13)

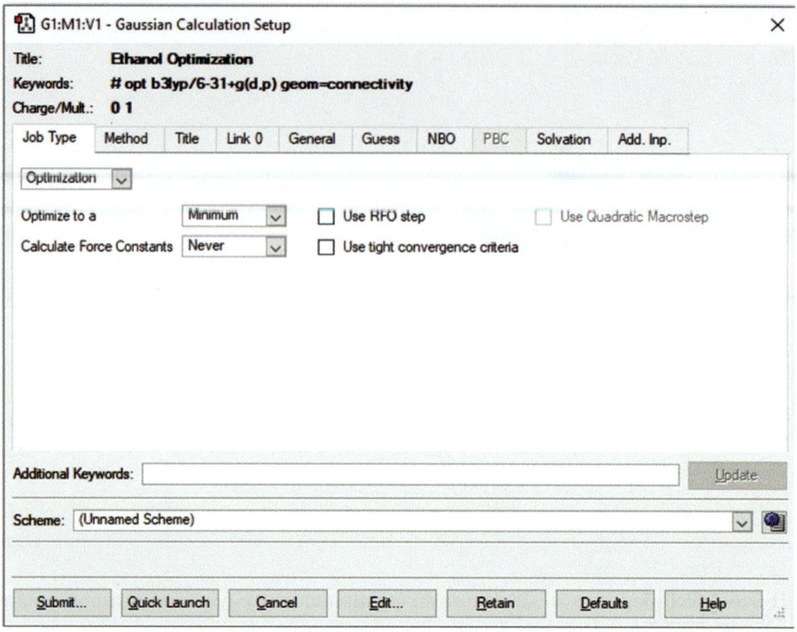

FIGURE 1.7.2 Interface of GaussView showing the selection of 'Optimization' as the job type for a Gaussian calculation.

Output from the geometry optimization using the .log file

Head Matter: Initial lines get printed at the top of every output file you generate. It contains information about the program, the company, and citation data.

Program Version: Lines below show the program version that was used for the computation and date it was run. Gaussian 09 Revision D.01 was used. This is important to track when reporting computational results in a scientific paper. Notice how Revision D.01 was released on April 13, 2013.

```
******************************************
Gaussian 09:  EM64W-G09RevD.01 13-Apr-2013
               27-Jan-2024
******************************************
```

Route Line: shows the route line that was given in the input file.

```
------------------------------------------
# opt b3lyp/6-31++g(d,p) geom=connectivity
------------------------------------------
```

Step 9: Press Ctrl+F, Search for "SCF Done" (Figure 1.14)

1.2.4 Frequency Analysis of Ethanol Molecule

Using the optimized geometry of ethanol molecule, the next step is to perform frequency analysis (Figures 1.15–1.18).

Step 1: Select Frequency as the job type.

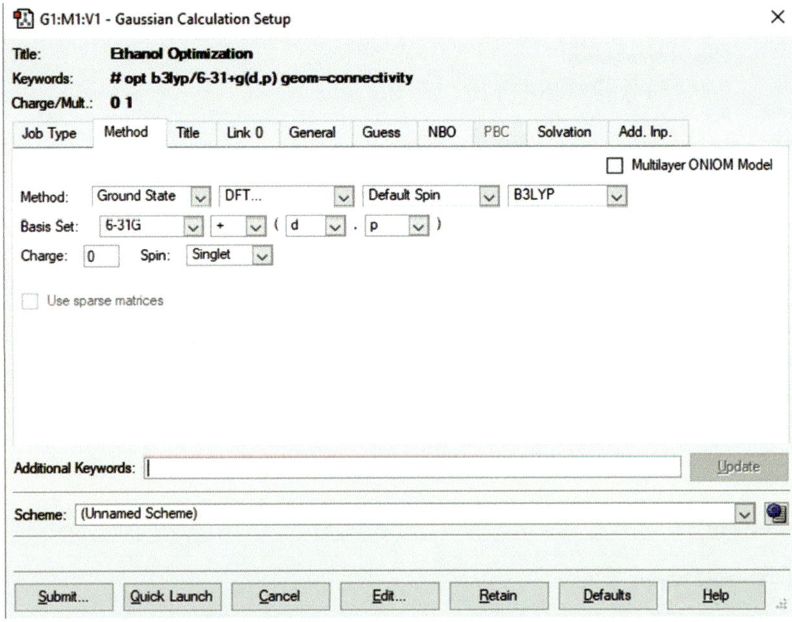

FIGURE 1.8 Screen capture of method selection in GaussView for geometry optimization, highlighting available theoretical methods and basis sets.

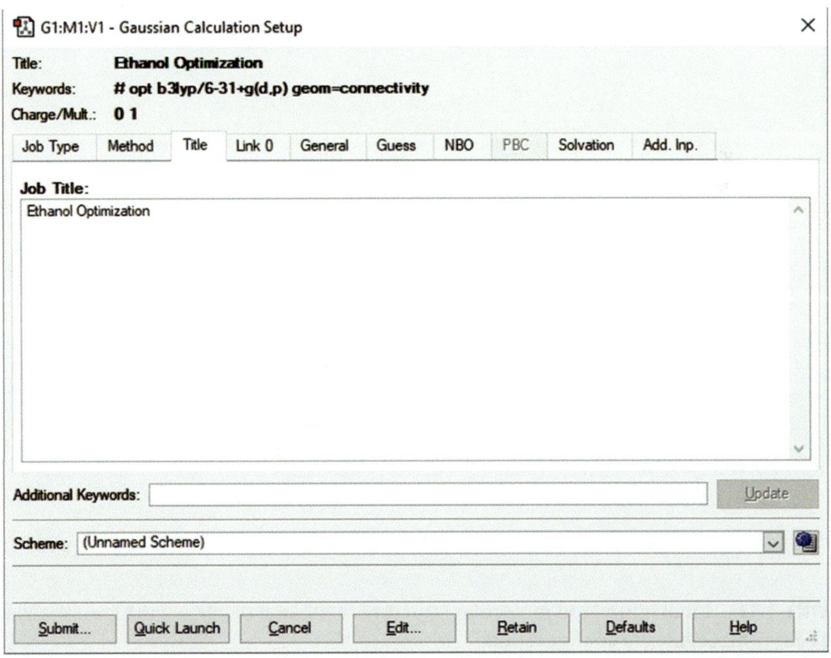

FIGURE 1.9 Interface snippet of GaussView showing where users can enter a title for their Gaussian job.

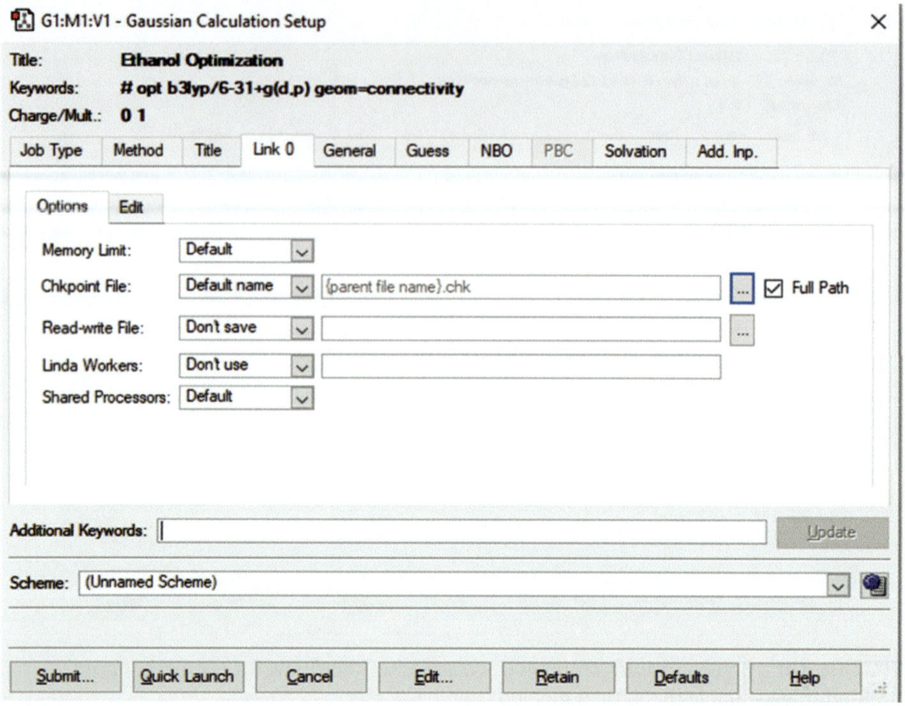

FIGURE 1.10 Display of the edit options available in GaussView, related to checkpoint file location, memory limit, and processors.

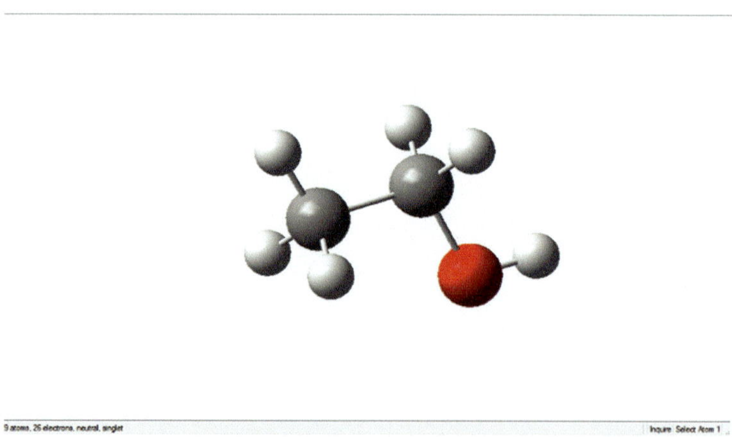

FIGURE 1.11 Final optimized molecular structure of ethanol displayed in GaussView, indicating minimized energy configuration.

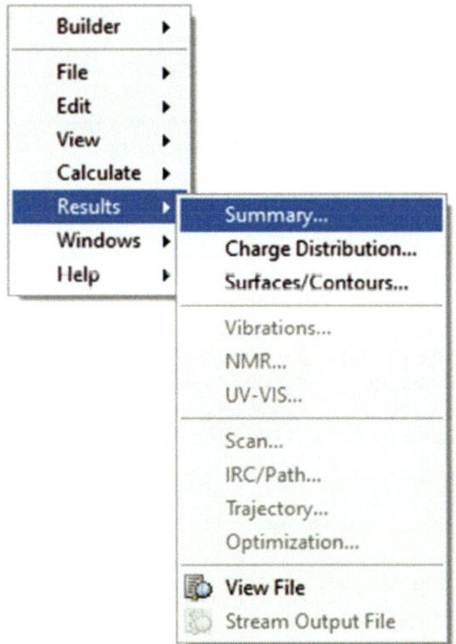

FIGURE 1.12 Visualization of molecular geometry and energy data for ethanol post-optimization in GaussView.

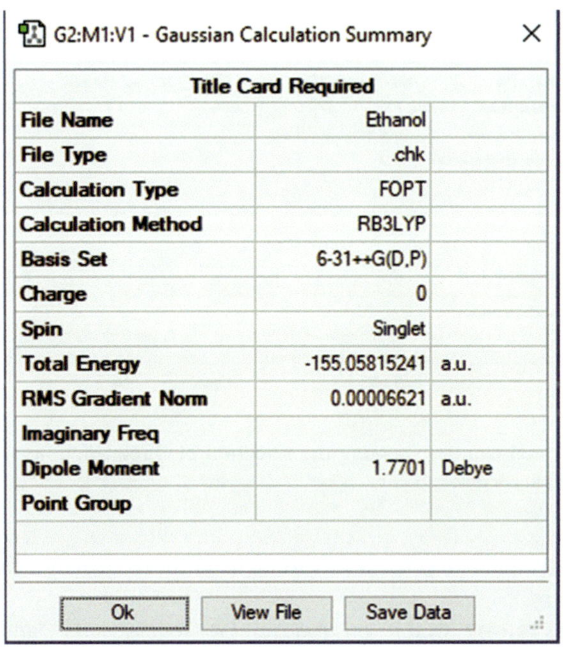

FIGURE 1.13 Summary tab interface in GaussView showing detailed computational results for a geometry optimized molecule.

```
SCF Done:  E(RB3LYP) =  -155.058152408     A.U. after    7 cycles
           NFock=   7  Conv=0.40D-08     -V/T= 2.0097
Calling FoFJK, ICntrl=      2127 FMM=F ISym2X=1 I1Cent= 0 IOpC1X= 0 NMat=1 NMatS=1 NMatT=0.
*** Axes restored to original set *****
-------------------------------------------------------------------
Center     Atomic                   Forces (Hartrees/Bohr)
Number     Number              X              Y              Z
-------------------------------------------------------------------
     1          6          0.000158367    0.000070116    0.000121418
     2          1         -0.000028941    0.000060612    0.000001008
     3          1         -0.000028930   -0.000031187   -0.000051989
     4          1          0.000083590    0.000007561    0.000013114
     5          6          0.000028124    0.000043346   -0.000075076
     6          1          0.000026595    0.000033621    0.000075050
     7          1          0.000026609   -0.000081800    0.000008415
     8          8         -0.000000506    0.000083344   -0.000144366
     9          1          0.000051826   -0.000030259    0.000052424
-------------------------------------------------------------------
Cartesian Forces:  Max     0.000158367 RMS     0.000066211
```

FIGURE 1.14 Display of a log file summary in GaussView, detailing the final optimized molecular geometry and associated computational parameters.

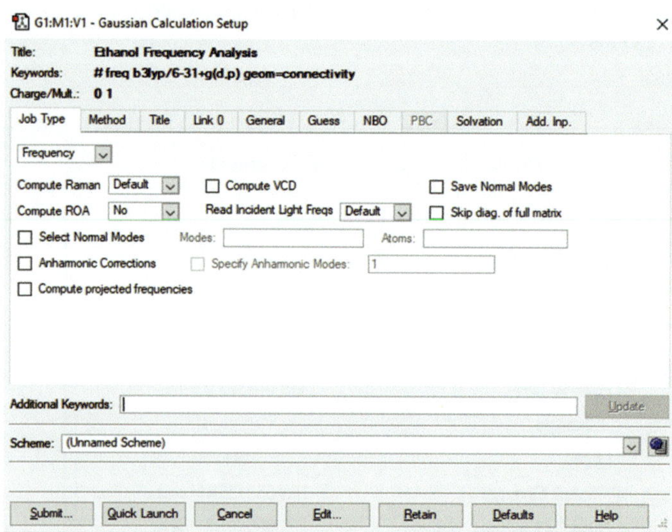

FIGURE 1.15 Screen capture showing the selection of 'Frequency' as job type for vibrational analysis in GaussView.

Step 2: Select method and basis sets to be used for frequency analysis of ethanol geometry.

Step 3: Press RIGHT CLICK on mouse, Go to "Results" and then select "Vibrations"

The display vibrations dialog displays the vibration that is presently chosen for dynamic display. Clicking the Start Animation button will start this display; the Stop Animation button will let you stop it. The molecule will start a cyclical displacement

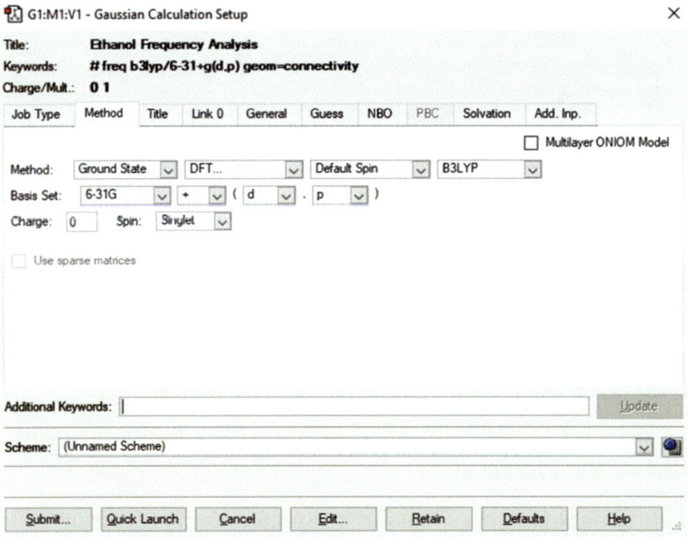

FIGURE 1.16 Duplicate of Figure 1–15, depicting the interface for setting up frequency analysis in GaussView.

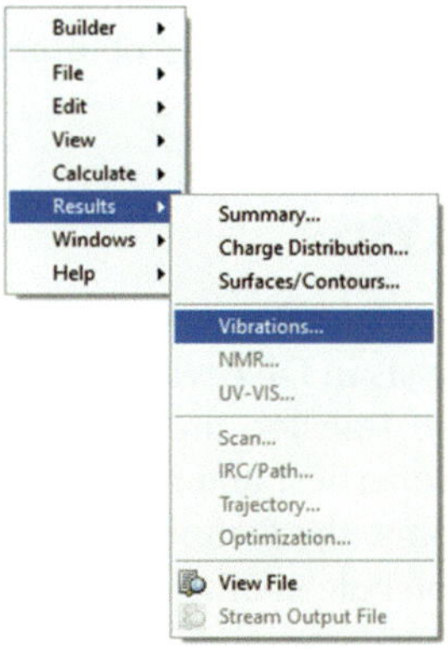

FIGURE 1.17 Results display in GaussView after completing frequency analysis, showing vibrational frequencies and related data.

displaying the motions matching the selected vibration. The two sliding bars beneath the START/STOP ANIMATION button change the motion's pace and scale. The vibrations preferences box allows one to specify default values for these variables as well as for the displacement and dipole derivative unit vector covered below.

The SAVE MOVIE option allows you to save the animation as either a movie or as individual frames. Choose another mode by choosing the desired one from the scrolling list. One can choose a new mode without stopping the old one. With the ordinary mouse buttons, you can also turn and move the vibrating molecule.

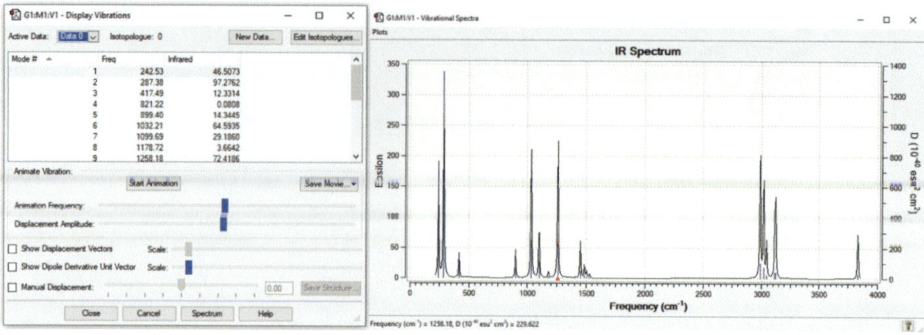

FIGURE 1.18 Interface showing the vibrational analysis and infrared spectrum of an ethanol molecule in GaussView.

1.2.5 Display the Spectrum

The spectrum symbol presents Gaussian's estimate of the spectrum in line format. Please be advised that the intensity values are relative to the maximum value within the current dataset and do not correspond directly to the experimental band intensities. Selecting the currently animated vibration mode can also be accomplished by clicking on the various peaks within the spectra.

Note: The other way to analyze or obtain the vibrational frequency data with other information in detail is to use .log file obtained after performing frequency analysis.

1.3 POPULATION ANALYSIS

Understanding electron distribution within a molecule requires the fundamental computational tool population analysis in DFT with Gaussian 09. Crucial for understanding chemical reactivity, charge transfer, and molecular interactions is knowledge of atomic charges, bonding properties, and molecule electrostatics. By distributing the electron density across atoms, the technique enables researchers to find partial atomic charges influencing molecule characteristics, dipole moments, and intermolecular forces. Gaussian 09 offers a number of population analysis techniques with differing degrees of basis-set dependency and precision.

Mulliken Population Analysis—which assigns electron density depending on atomic orbital coefficients—is among the most often utilized techniques. Though simple, Mulliken charges are quite sensitive to the foundation set choice, which often results in conflicts. Using orthogonalized AOs reduces basis-set dependency while still depending on arbitrary charge partitioning, hence providing an improvement over Lowdin population analysis. Natural population analysis (NPA) using the natural bond orbital (NBO) technique offers a more chemically significant methodology that offers a strong description of charge distribution with low basis-set dependence. Comparatively to Mulliken and Lowdin approaches, another often used method is the Hirshfeld method, which assigns atomic charges based on the electron density of unbound atoms, hence more physically accurate.

Commonly used for applications in molecular dynamics and force-field creation are electrostatic potential (ESP)-derived charges like Merz-Kollman (MK) and CHELPG charges. These techniques guarantee correct modeling of intermolecular interactions by matching atomic charges to mimic the molecular electrostatic potential. Development of polarizable force fields and parameterizing molecular mechanics models benefits especially from ESP-derived charges. The intended use determines the population analysis technique; NPA is favored for chemical bond ing investigations and ESP-derived charges for simulations needing exact electrostatic interactions. Though useful, all population analysis methods have inherent approximations; therefore, results should be understood in context with other computational and experimental data to get a complete knowledge of electronic structure.

1.3.1 Example: Population Analysis on Ethanol Molecule

Step 1: Use the .log file from geometry optimization or frequency analysis and search for "Mulliken", which is default used by Gaussian package.

Step 2: While selecting method and basis sets for optimization of geometry, enter a keyword "POP=NBO" for performing NBO charge analysis. Users can also use other population analysis as per their need whose details can be found Gaussian software manual (Figures 1.19 and 1.20).

1.4 THERMOCHEMISTRY ANALYSIS

DFT is a significant quantum mechanical modeling method utilized in computational chemistry to examine the electronic structure of atoms and molecules. Thermochemistry is the specific discipline of chemistry that investigates heat changes in chemical reactions. Integrating DFT with thermochemical analysis enables researchers to gain deep insights into the thermodynamic properties of molecules and processes. The precision of thermochemical predictions is significantly affected by the choice of exchange-correlation functional and basis set, despite DFT being a powerful method for investigating molecular properties.

Note: Frequency calculations yield thermochemical data that may be accessed in detail through the .log file for frequency analysis from the Gaussian software

```
Mulliken charges:
               1
     1  C   -0.490006
     2  H    0.171945
     3  H    0.171945
     4  H    0.132333
     5  C   -0.101881
     6  H    0.127021
     7  H    0.127021
     8  O   -0.497485
     9  H    0.359106
Sum of Mulliken charges =   0.00000
```

FIGURE 1.19 Visualization of Mulliken atomic charges on an ethanol molecule in GaussView, indicating electron density distribution.

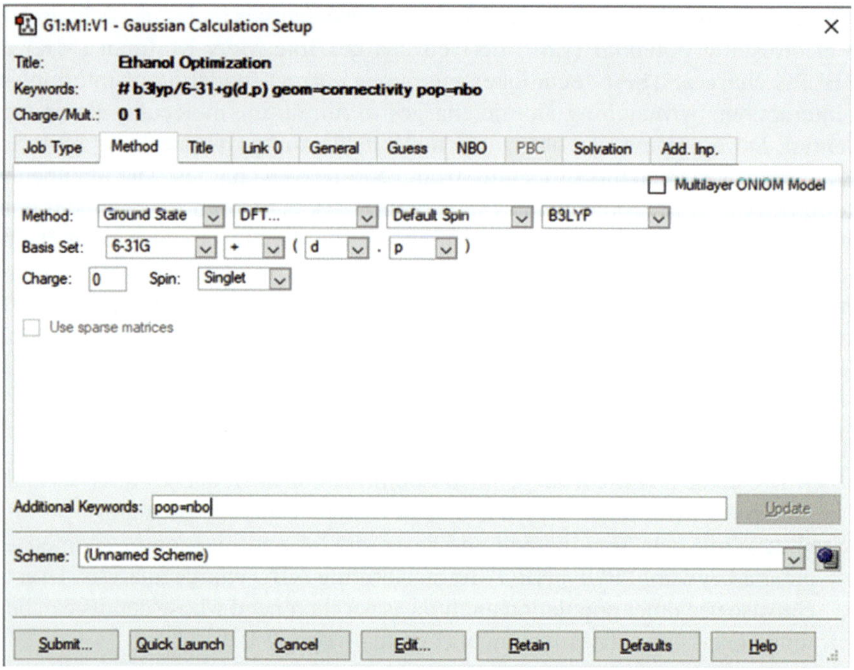

FIGURE 1.20 Setup for NBO charge analysis in GaussView, detailing the method and basis set selection for ethanol optimization.

package. The specifics are outlined in Figure 1.21 for an ethanol molecule. A comprehensive report on these data is available here. (https://gaussian.com/thermo/)

1.5 SPECTROSCOPY ANALYSIS

Two crucial methods for figuring out a molecule's structure are spectroscopic analysis and theoretical computational investigations. Electronic spectra are based on a variety of physical and chemical molecular characteristics. Modifying the spectrum properties of molecules allows for the easy computational investigation of various chemical and physical impacts, as many exciting chemical problems involve both ground and excited states of molecules [2].

DFT is occasionally used to compute a molecule's different thermodynamic properties and is employed in physics and chemistry to study electronic structure, primarily in the ground state. These systems emit a wide range of wavelengths, which allows for a thorough investigation of their structures, including the ground state and excited state electron configurations [3].

1.5.1 UV-Visible Spectra

The ability of DFT to predict electronic excitation energies makes it possible to simulate UV–Vis spectra. This facilitates understanding of the electronic structure

```
-------------------
- Thermochemistry -
-------------------
Temperature 298.150 Kelvin. Pressure 1.00000 Atm.
Atom   1 has atomic number 6 and mass  12.00000
Atom   2 has atomic number 1 and mass   1.00783
Atom   3 has atomic number 1 and mass   1.00783
Atom   4 has atomic number 1 and mass   1.00783
Atom   5 has atomic number 6 and mass  12.00000
Atom   6 has atomic number 1 and mass   1.00783
Atom   7 has atomic number 1 and mass   1.00783
Atom   8 has atomic number 8 and mass  15.99491
Atom   9 has atomic number 1 and mass   1.00783
Molecular mass:   46.04186 amu.
Principal axes and moments of inertia in atomic units:
                          1          2          3
    Eigenvalues --   51.56668  194.01220  222.86303
          X           0.99445    0.10519    0.00000
          Y          -0.10519    0.99445    0.00000
          Z           0.00000    0.00000    1.00000
This molecule is an asymmetric top.
Rotational symmetry number 1.
Rotational temperatures (Kelvin)      1.67965     0.44643     0.38864
Rotational constants (GHZ):          34.99820     9.30220     8.09798
Zero-point vibrational energy    209577.2 (Joules/Mol)
                                   50.09016 (Kcal/Mol)
Warning -- explicit consideration of 3 degrees of freedom as
           vibrations may cause significant error
Vibrational temperatures: 348.95   413.47   600.68  1181.55  1294.03
         (Kelvin)        1485.12  1582.21  1695.91  1810.24  1867.06
                         2022.14  2087.38  2139.69  2164.67  2200.70
                         4300.64  4343.58  4380.42  4483.54  4494.18
                         5516.47

Zero-point correction=                           0.079824 (Hartree/Particle)
Thermal correction to Energy=                    0.084137
Thermal correction to Enthalpy=                  0.085081
Thermal correction to Gibbs Free Energy-         0.054409
Sum of electronic and zero-point Energies=         -154.978329
Sum of electronic and thermal Energies=            -154.974016
Sum of electronic and thermal Enthalpies=          -154.973071
```

FIGURE 1.21 Detailed output of thermochemical data for an ethanol molecule, displaying zeropoint energy, enthalpy, and Gibbs free energy corrections in Hartrees.

of molecules and how they absorb light. UV-Visible (UV–Vis) spectra of molecules are frequently predicted using DFT, which offers important insights into electronic transitions and absorption characteristics. Steps involved in how UV–Vis spectra are simulated using DFT are as follows.

1.5.1.1 Geometry Optimization: Input File Preparation for UV-Spectra Analysis

Let's take the example of the ethanol molecule as performed in previous sections.

In molecular optimization, iterative modifications are performed on the molecular structure to ascertain the lowest energy configuration. The convergence criteria are crucial for determining the precise moment at which the optimization process attains an acceptable degree of precision. These criteria guarantee that the computations yield a robust and dependable result. Here are some of the often employed convergence criteria in the domain of geometry optimization.

The wavelengths of absorption and emission of chemical substances can be ascertained using time-dependent DFT (TD-DFT). UV–Vis spectroscopy is a quantitative technique for measuring the light absorption of a substance. This is achieved by contrasting the light intensity transmitted through a sample with that of a reference sample.

To calculate the UV-spectra of ethanol, the molecule was optimized with B3LYP/6–31G model chemistry (Figures 1.22 and 1.23).

After optimization is done, the next step can be used to view the summary of the optimized outcomes.

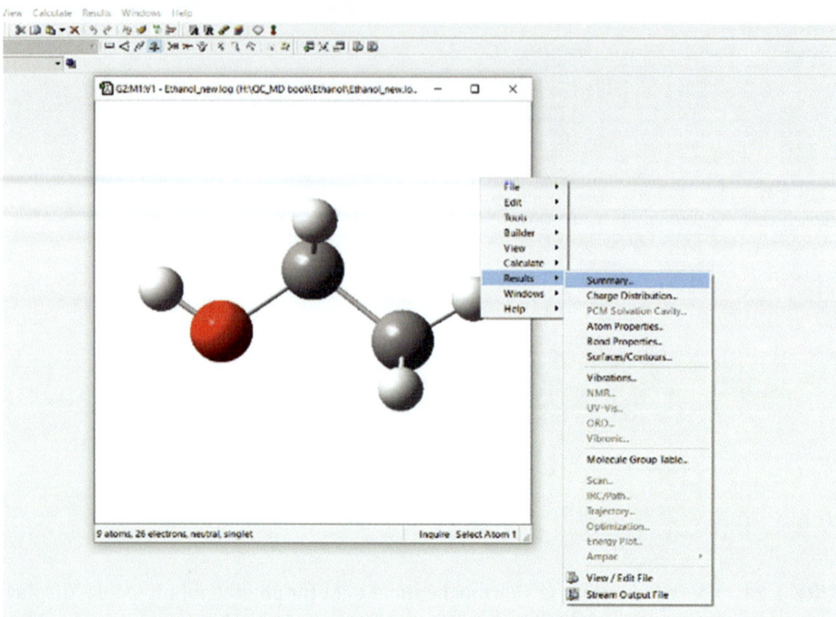

FIGURE 1.22 Interface of the GaussView showing the tab to set up UV-Vis spectral analysis for an optimized ethanol molecule.

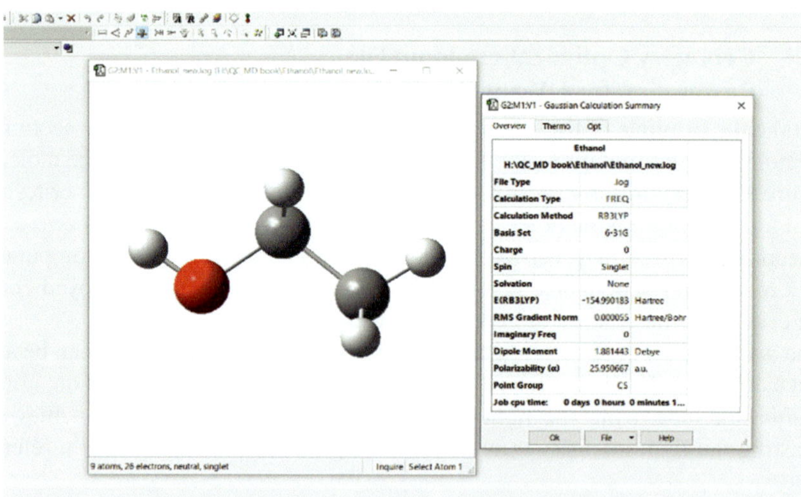

FIGURE 1.23 Summary output screen displaying the results of geometry optimization for an ethanol molecule, including key molecular parameters.

FIGURE 1.24 Screenshot of GaussView interface detailing the steps to configure a single point energy calculation using time-dependent DFT for ethanol.

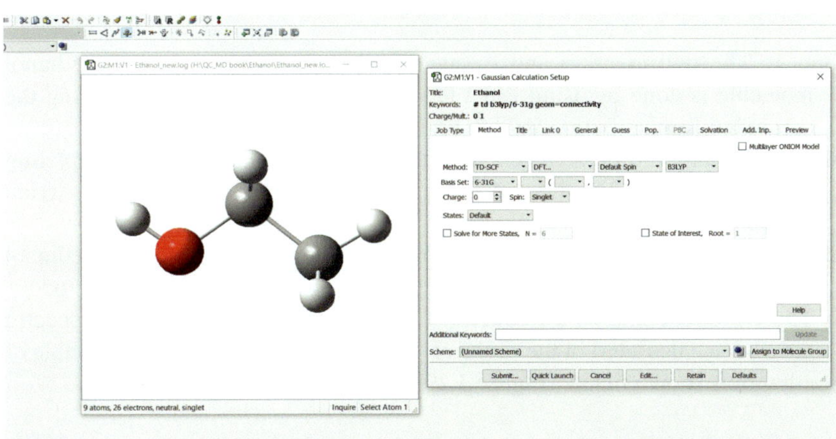

FIGURE 1.25 Interface showing the setup process for a TD-SCF calculation on an ethanol molecule, focusing on method and basis set selections.

Thereafter, a single-point calculation has been done through the following step, keeping TD-SCF in the method section and the rest are the same (Figures 1.24 and 1.25).

After that, the job has been submitted and the calculations have run. After successfully completing the job, go to the analysis part (Figure 1.26).

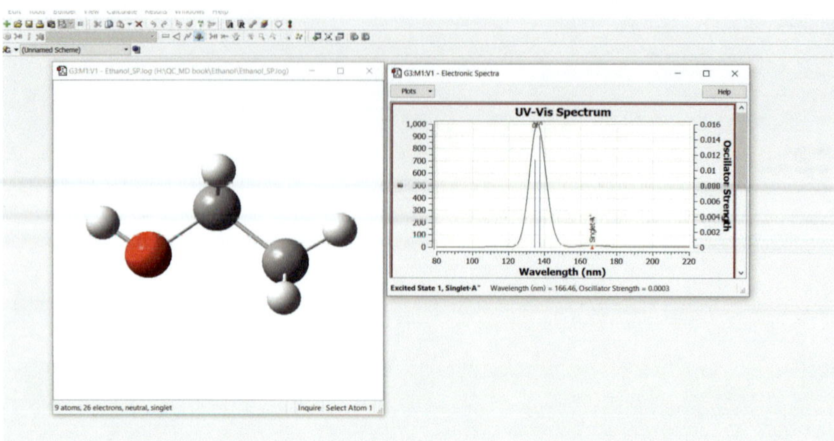

FIGURE 1.26 Graphical display of the UV-Vis absorption spectrum of ethanol, showcasing peaks corresponding to different electronic transitions.

1.5.2 Raman Spectra

As a novel kind of nondestructive assessment technology, Raman spectroscopy is one of the most significant techniques for the spectroscopic analysis of substances and is frequently used for both qualitative and quantitative analysis of samples. Raman spectroscopy calculation of ethanol was shown in the following steps:

Step 1: The optimization and frequency analysis (opt+freq) of the ethanol molecule is done by using B3LYP/6–31G level of theory and choosing the computed Raman option in "job type section" (Figure 1.27)

Step 2: Submit the desired job in Gaussian. After successfully completing the job, open the .log file and go to vibration analysis in the result section (Figure 1.28).

Step 3: Open the vibration part, where the different frequencies of ethanol can be seen. Click the spectrum button in the display vibration window. The window contains a frequency plot of Raman activity. Raman spectra have been calculated in the range of 400–4000 cm^{-1}. The frequency table of ethanol is shown below (Figure 1.29)

The spectra of the following molecules are shown below

The basis of Raman spectroscopy is the inelastic scattering of light in or near the visible spectrum, which happens when light interacts with molecular vibrations and causes characteristic up- or downshifts in the applied photons' energy. The Raman activity is the quantity of scattering (Figure 1.30).

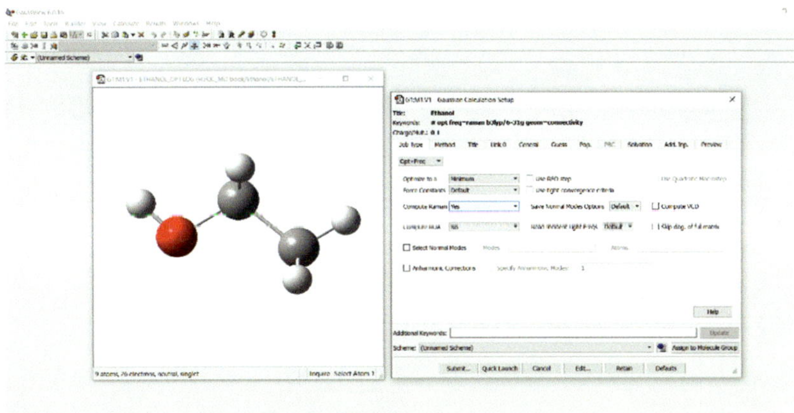

FIGURE 1.27 Screen capture of the GaussView setup dialog for performing a Raman frequency analysis on an ethanol molecule.

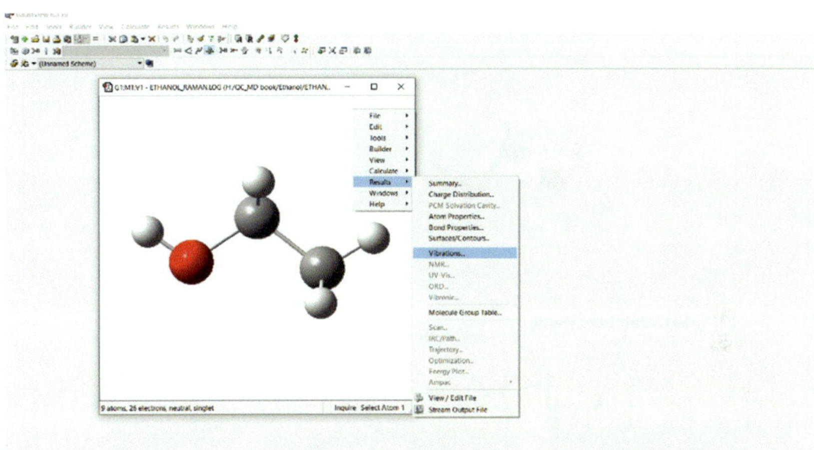

FIGURE 1.28 Detailed view of the GaussView interface used to access and visualize vibrational data from Raman spectroscopy analysis.

1.5.3 IR Spectra

Similar step has been followed to calculate IR spectra and also discussed in Section 1.2 for performing frequency analysis. After successfully completing the frequency analysis, click the left corner of the vibrational spectra window and unclick the Raman spectra (Figure 1.31).

The plot of IR spectra has been shown with frequency data (Figure 1.32).

The spectrum shows one strong peak and four weak peaks.

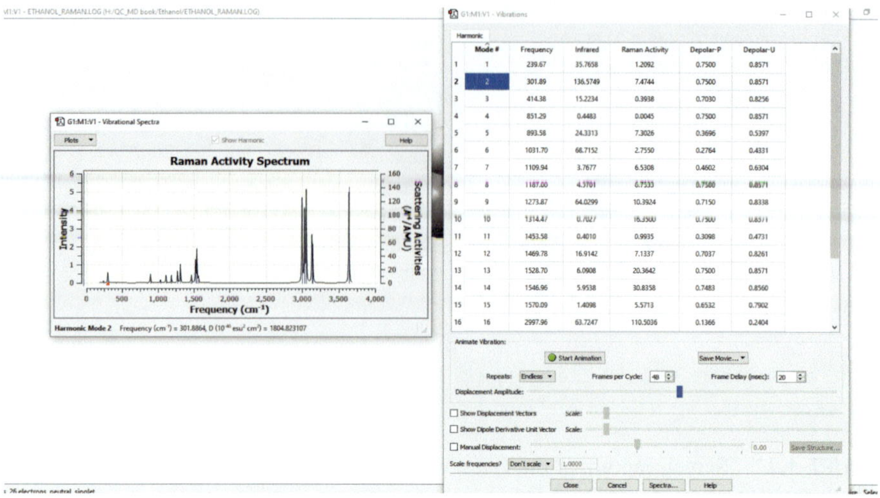

FIGURE 1.29 Display of Raman activity spectrum for ethanol showing various vibrational modes and their corresponding frequencies.

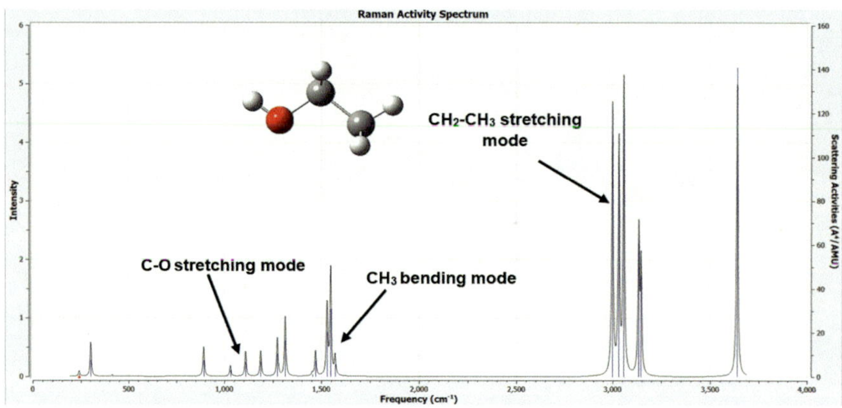

FIGURE 1.30 Visual representation of theoretical Raman spectrum highlighting the vibrational frequencies and Raman activity of an ethanol molecule.

1.5.4 NMR

In chemistry and biology, nuclear magnetic resonance (NMR) is a potent analytical method used to examine the dynamics, composition, and structure of molecules. The fundamental concept underlying NMR is that certain atomic nuclei possess distinct spin states when subjected to an external magnetic field. NMR detects transitions between spin states that are unique to the precise nuclei being studied, as well as the chemical surroundings of those nuclei.

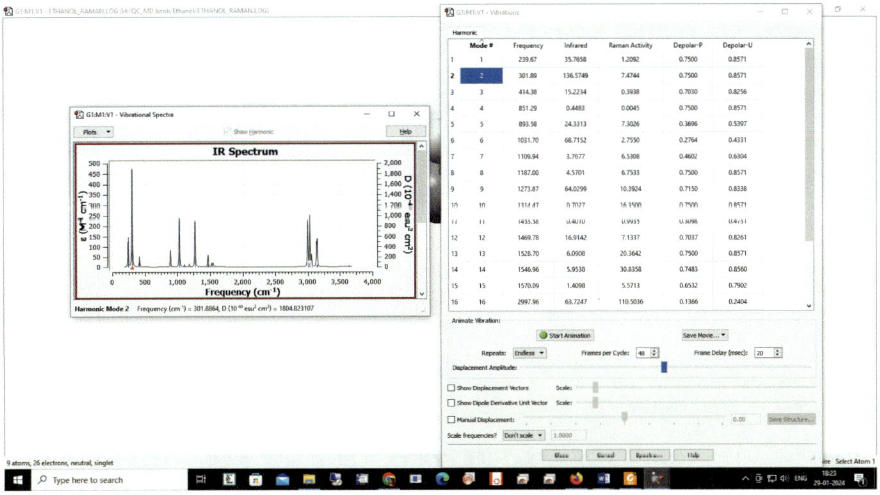

FIGURE 1.31 Graph showing the infrared spectrum of ethanol with peaks labeled with their respective vibrational frequencies.

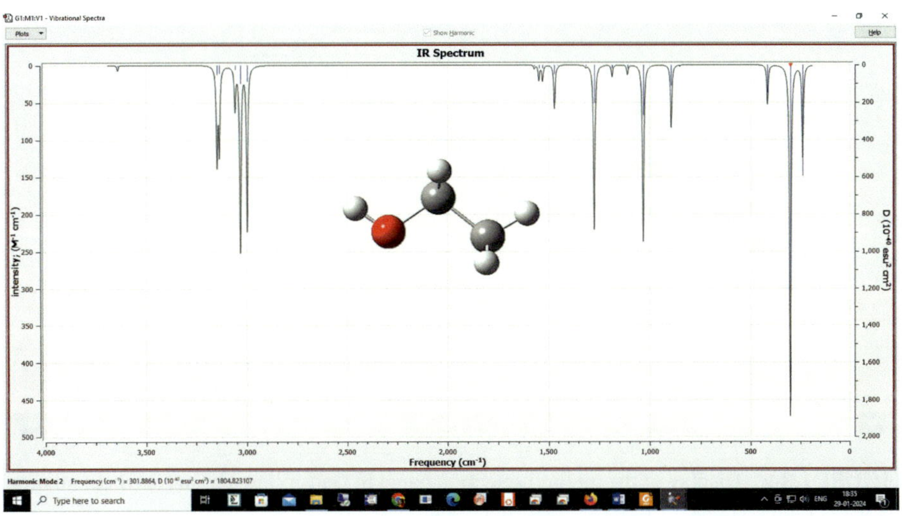

FIGURE 1.32 Detailed infrared spectrum of ethanol showing a strong peak and several weaker peaks, each annotated with vibrational frequency data.

The basis of NMR is the interaction of an external magnetic field with the magnetic moments of atomic nuclei. Some nuclei absorb energy and change from a lower energy state to a higher energy state when they are exposed to a strong magnetic field and radiofrequency (RF) pulse.

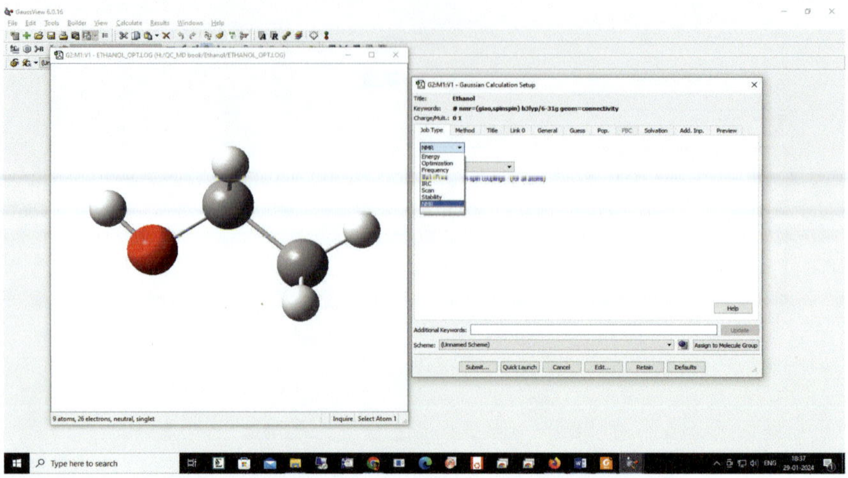

FIGURE 1.33 Interface screenshot showing the setup options in GaussView for performing NMRcalculations on an ethanol molecule.

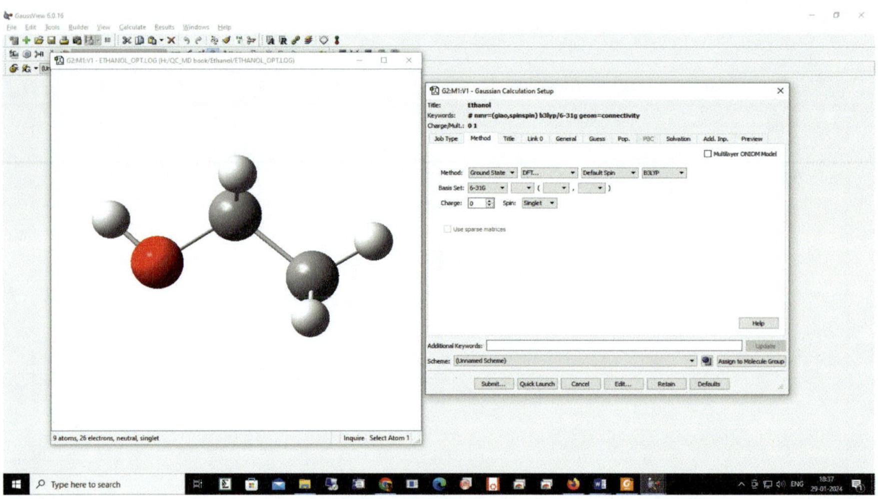

FIGURE 1.34 Display of the method selection panel in GaussView for an NMR calculation, focusing on the choice of basis set and functional.

As an example, the ethanol molecule is considered for the calculation of NMR in Gaussian.

Step 1: In the job type section in Gauss view, select the NMR option (Figure 1.33)

Step 2: Thereafter, in the method section, choose ground state and select 6–31G basis set and B3LYP level of DFT and submit the job (Figure 1.34).

Step 3: After completing the desired job, open NMR in the result section (Figures 1.35 and 1.36).

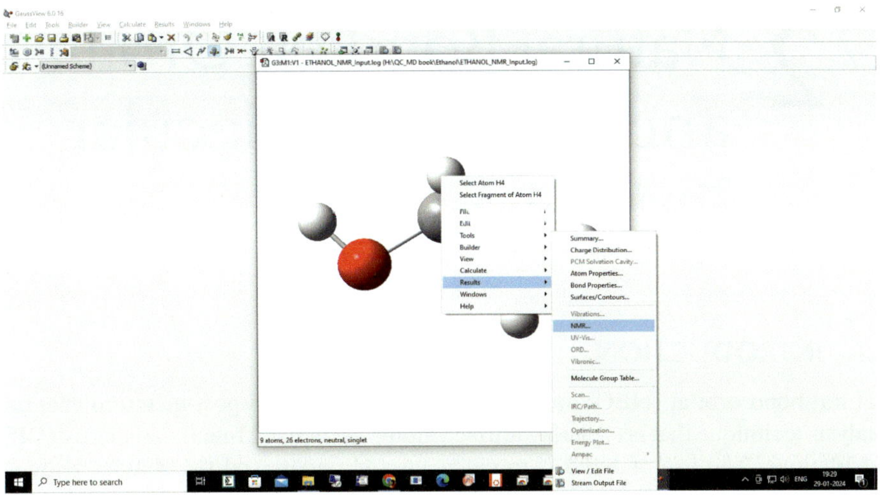

FIGURE 1.35 Step-by-step guide shown in GaussView for accessing and analyzing NMR data from a completed calculation.

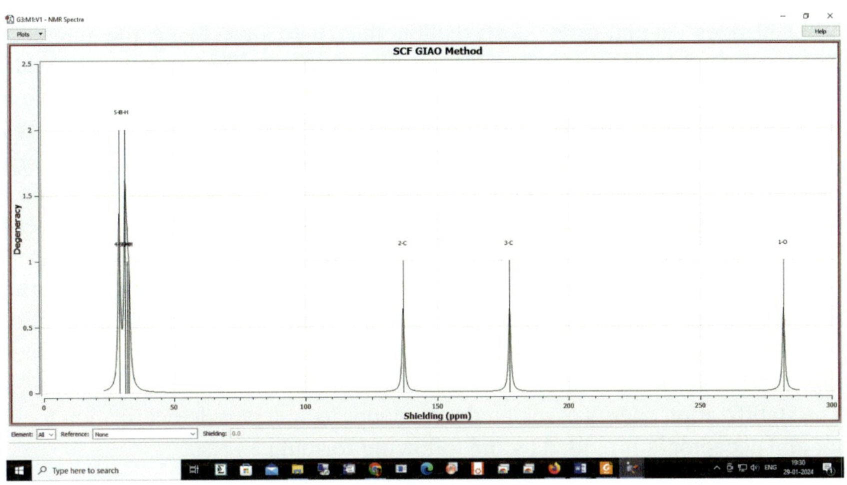

FIGURE 1.36 Theoretical NMR spectrum of an ethanol molecule, displaying peaks that correspond to different nuclear environments within the molecule.

REFERENCES

1. Jensen, F. (2017). *Introduction to Computational Chemistry.* United Kingdom: Wiley.
2. Altaf, A.A., S. Kausar, and A. Badshah, Spectral calculations with DFT. *Density Functional Calculations: Recent Progresses of Theory and Application*, 2018. **101**.
3. Fathima Rizwana, B., J.C. Prasana, and S. Muthu, Spectroscopic investigation (FT-IR, FT-Raman, UV, NMR), Computational analysis (DFT method) and Molecular docking studies on 2-[(acetyloxy) methyl] -4-(2-amino-9h-purin-9-yl) butyl acetate. *International Journal of Materials Science*, 2017. **12**: p. 196–216.

2 Natural Bonding Orbital and Frontier Molecular Orbital Analysis

2.1 INTRODUCTION

Natural bond orbital (NBO) analysis is a widely employed post-quantum chemical analysis technique that is typically utilized alongside density functional theory (DFT) calculations. The use of NBO is to calculate and understand the characteristics of chemical bonding inside molecular systems. It provides a way to understand bonding patterns, electron delocalization, and the nature of interactions within a molecule. NBO was developed upon the principles of the natural orbital theory, which seeks to depict the electronic configuration of a molecule in a manner that highlights the presence of localized bonds [1].

NBO utilizes second-order perturbation theory to investigate the interactions between pairs of localized molecular orbitals (LMOs). The interactions and the energy associated with interaction can be calculated by NBO analysis with the help of perturbation theory. In the optimal Lewis structure, the localized orbitals can interact significantly. An empty or filled bonding, antibonding, or lone pair orbital can function as an acceptor, and a full bonding or lone pair orbital can function as a donor. Bonds can be strengthened or weakened by these exchanges. Donor–acceptor charge transfer, or $\sigma \rightarrow \sigma^*$ interactions between filled (donor) and unfilled (acceptor) orbitals, is linked to the non-covalent delocalization effects [2]. NBO analysis offers a practical foundation for studying charge transfer or conjugative interaction in molecular systems, as well as an effective way to examine intra- and intermolecular bonding and bond interaction.

The stabilization energy associated with donor to acceptor delocalization for each donor and acceptor is provided by the following formula.

$$\Delta E_{ij} = q_i \frac{F^2(i,j)}{\varepsilon_j - \varepsilon_i} \tag{2.1}$$

Here, *i* represents donor of NBO and *j* represent acceptor of NBO. *i*th donor occupancy is denoted by q_i, whereas the diagonal elements (orbital energies) associated with the NBO fock matrix [3] are represented by ε_i, ε_j. The off-diagonal elements are represented by $F(i, j)$, respectively.

DOI: 10.1201/9781032694733-2

2.2 NBO CHARGE ANALYSIS

To calculate NBO, let us consider an example of lutidine, which is a dimethyl substitute derivative of pyridine. The lutidine structure, as shown in Figure 2.1, was optimized at the B3LYP level of theory with a basis set 6-31G.

The optimized structure of the lutidine with atomic number is shown in Figure 2.2

The keywords in order to do the NBO analysis include the "Population=NBO" option. Additionally, we need to specify another option to save the NBOs in the

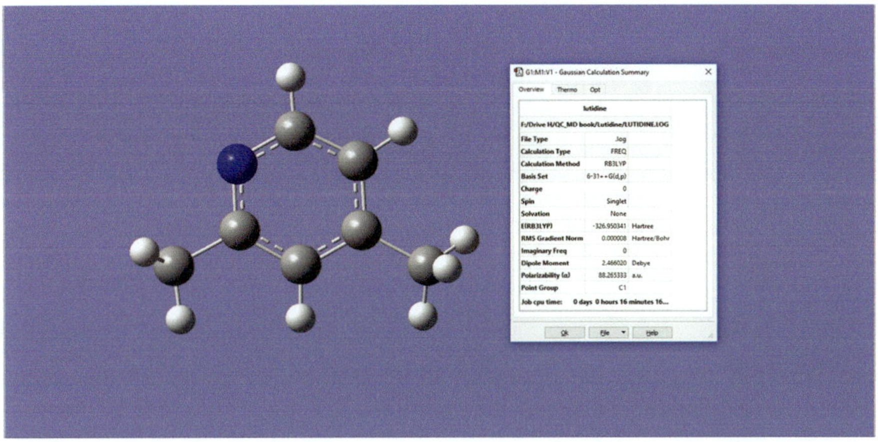

FIGURE 2.1 The optimized Lutidine molecule with respective calculation result summary.

FIGURE 2.2 The optimized structure of the Lutidine with atomic notations.

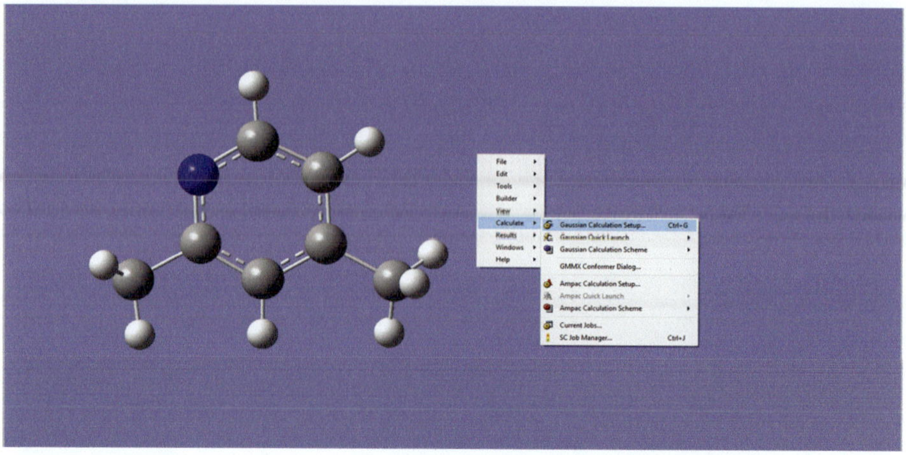

FIGURE 2.3 The steps to select the Gaussian calculation setup in the Gaussian module.

checkpoint file. Or, once the optimization is done, we can move to the calculation part by opening the log file of the optimized structure. The steps are shown in Figures 2.3 and 2.4.

Next step, in calculation window, we select energy in job type.

In the Method section, the other options do remain the same as shown in Figure 2.5.

Thereafter, we go to the pop section, click perform NBO calculation, and choose the full NBO. After that click save orbitals to .chk file and choose save NBOs. Submit the job. Steps are displayed in Figures 2.6 and 2.7.

After successful completion of the job, we open the log file in notepad. The following output is obtained, as shown in Figure 2.8.

The output of NBO first presents a summary of the findings from natural population analysis (NPA). The primary natural atomic orbitals (NAO) table is shown in the first section. In Figure 2.9, there are a total of 239 NAO functions that are reported.

The information includes the angular momentum type (s, p_x, p_y, p_z, etc.), the orbital type (core, valence, or Rydberg), the orbital occupancy (number of electrons or "natural population" of the orbital), and the orbital energy (in atomic units: 1 a.u. = 627.5 kcal/mol). It is important to observe that the Rydberg (Ryd) NAOs generally have significantly lower occupancies compared to the core (Cor) and valence (Val) NAOs of the natural minimum basis (NMB) set. This difference in occupancy levels highlights the primary importance of the NMB orbitals in accurately representing molecular characteristics. If a Fock matrix is available, the primary quantum numbers for the NAO labels (1s, 2s, 3s, etc.) are assigned based on the energy order; if not, they are assigned based on occupancy [1]. If the occupancy and energy ordering do not match, a message notifying user of a "population inversion" is printed.

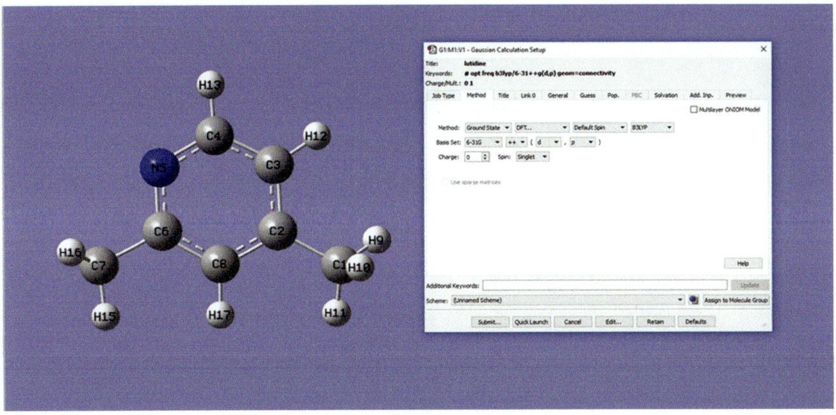

FIGURE 2.4 The optimized structure of the Lutidine with the respective Gaussian calculation setup window.

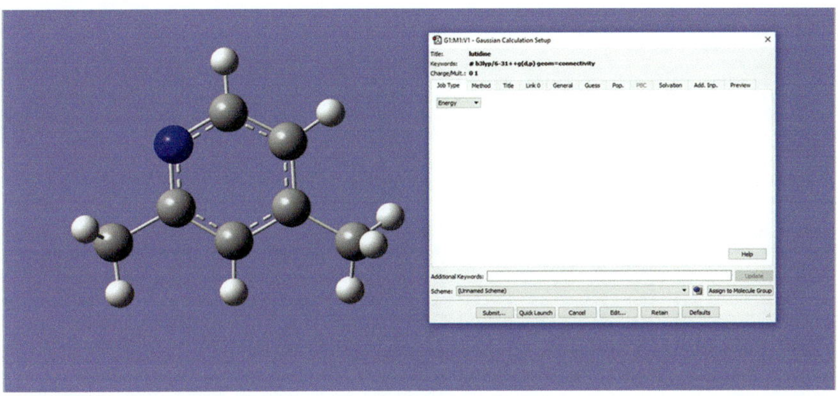

FIGURE 2.5 The optimized structure of the Lutidine with respective method tab details in gaussian calculation setup window.

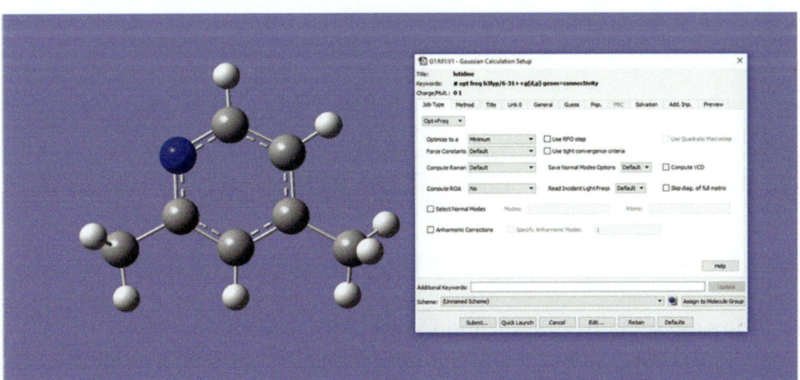

FIGURE 2.6 The optimized structure of the Lutidine with the respective Gaussian calculation setup window with an arrow on NBO.

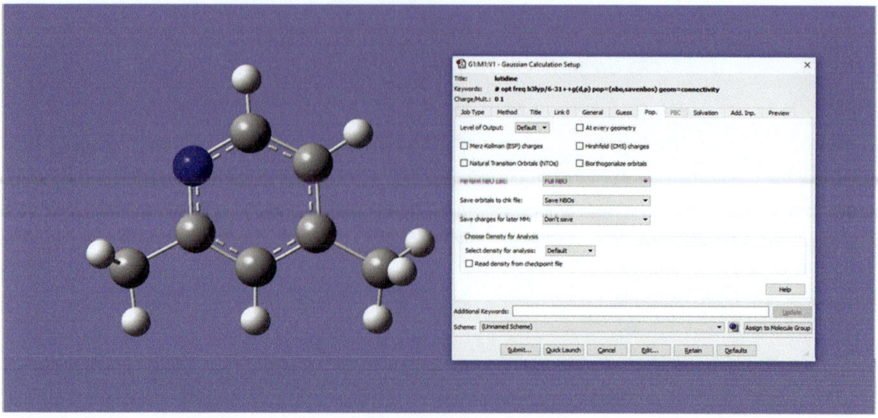

FIGURE 2.7 The optimized structure of the Lutidine with the respective Gaussian calculation setup with a circle on the Perform NBO calculation tab.

```
Sorting of NBOs:                 67   69   30   61   68   52   41   54   36   39
Sorting of NBOs:                 59   55   50   48   46   35   45   40   53   37
Sorting of NBOs:                 60   43   51   57   38   58   34   33   47   49
Reordering of NBOs for storage:  25   26   24   22   21   23   27   28   13   11
Reordering of NBOs for storage:   8   16    6    5    1   15   10   20   12    4
Reordering of NBOs for storage:   2    3   18   17   19   29   14    9    7   84
Reordering of NBOs for storage:  79   77   71   85   73   72   74   88   87   89
Reordering of NBOs for storage:  90   80   82   83   81   75   76   86   78   65
Reordering of NBOs for storage:  66   70   44   31   32   56   42   63   62   64
Reordering of NBOs for storage:  67   69   30   61   68   52   41   54   36   39
Reordering of NBOs for storage:  59   55   50   48   46   35   45   40   53   37
Reordering of NBOs for storage:  60   43   51   57   38   58   34   33   47   49
Labels of output orbitals:  CR  CR  CR  CR  CR  CR  CR  CR  BD  BD  BD  BD  BD  BD  BD  BD  BD  BD  BD  BD
Labels of output orbitals:  BD  BD  BD  BD  BD  LP  BD  BD  BD  BD* BD* BD* BD* BD* BD* BD* BD* BD* BD* BD*
Labels of output orbitals:  BD* BD* BD* BD* BD* BD* BD* BD* BD* RY* RY* RY* RY* RY* RY* RY* RY* RY* RY* RY*
Labels of output orbitals:  RY* RY* RY* RY* RY* RY* RY* RY* RY* RY* RY* RY* RY* RY* RY* RY* RY* RY* RY* RY*
Labels of output orbitals:  RY* RY* RY* RY* RY* RY* RY* RY* RY* RY*
1|1|UNPC-CLPC4|SP|RB3LYP|6-31G|C7H9N1|COMPUTER LAB|05-Feb-2024|0||# b3
lyp/6-31g pop=(nbo,savenbo) geom=connectivity||lutidine||0,1|C,0,2.551
582,-0.98886,0.007349|C,0,1.228128,-0.260572,-0.009778|C,0,1.166239,1.
14231,-0.006912|C,0,-0.081995,1.767999,0.002288|N,0,-1.251573,1.093854
,0.006554|C,0,-1.207759,-0.2629,0.000811|C,0,-2.534933,-0.976219,0.001
594|C,0,0.010424,-0.958585,-0.0087|H,0,3.301068,-0.469624,-0.599657|H,
0,2.94969,-1.055328,1.028811|H,0,2.451504,-2.010361,-0.372999|H,0,2.07
4575,1.736621,-0.01374|H,0,-0.158964,2.850997,0.003612|H,0,-3.122074,-
0.692088,-0.879157|H,0,-2.410836,-2.063178,0.003571|H,0,-3.12226,-0.68
855,0.880968|H,0,0.00819,-2.044506,-0.01719||Version=IA32W-G09RevC.01|
State=1-A|HF=-326.8378272|RMSD=3.626e-009|Dipole=0.8031689,-0.5779678,
0.0160081|Quadrupole=0.8863841,1.0870413,-1.9734254,2.7538851,0.104349
3,-0.0414846|PG=C01 [X(C7H9N1)]||@


ALTHOUGH J.J. COULD DIAGNOSE THE FAULTS OF AN
APPARATUS WITH UNCANNY ACCURACY, IT WAS JUST AS
WELL NOT TO LET HIM HANDLE IT.
  -- GEORGE THOMPSON, ABOUT HIS FATHER
Job cpu time:  0 days  0 hours  0 minutes 24.0 seconds.
File lengths (MBytes):  RWF=     10 Int=      0 D2E=      0 Chk=      1 Scr=      1
Normal termination of Gaussian 09 at Mon Feb 05 11:49:21 2024.
```

FIGURE 2.8 The output result of the optimized structure of the Lutidine.

The following section, as shown in Figure 2.10, presents an atomic summary that displays the total core, valence, and Rydberg populations on each atom as well as the natural atomic charges (nuclear charge minus summed natural populations of NAOs on the atom).

```
Analyzing the SCF density

Job title: lutidine

Storage needed:     25274 in NPA,     32968 in NBO (  33553511 available)


NATURAL POPULATIONS:  Natural atomic orbital occupancies
```

NAO	Atom	No	lang	Type(AO)	Occupancy	Energy
1	C	1	s	Cor(1S)	1.99924	-10.05536
2	C	1	s	Val(2S)	1.09442	-0.26042
3	C	1	s	Ryd(3S)	0.00023	1.27569
4	C	1	px	Val(2p)	1.13319	-0.08887
5	C	1	px	Ryd(3p)	0.00152	0.54081
6	C	1	py	Val(2p)	1.22569	-0.09771
7	C	1	py	Ryd(3p)	0.00247	0.62464
8	C	1	pz	Val(2p)	1.25830	-0.10127
9	C	1	pz	Ryd(3p)	0.00027	0.62902
10	C	2	s	Cor(1S)	1.99894	-10.06389
11	C	2	s	Val(2S)	0.88264	-0.15050
12	C	2	s	Ryd(3S)	0.00058	0.98507
13	C	2	px	Val(2p)	1.07430	-0.06190
14	C	2	px	Ryd(3p)	0.00486	1.08441
15	C	2	py	Val(2p)	1.08773	-0.06092
16	C	2	py	Ryd(3p)	0.00447	0.94040
17	C	2	pz	Val(2p)	0.94195	-0.10726
18	C	2	pz	Ryd(3p)	0.00140	0.67671
19	C	3	s	Cor(1S)	1.99891	-10.04224
20	C	3	s	Val(2S)	0.96055	-0.16656
21	C	3	s	Ryd(3S)	0.00085	0.99909
22	C	3	px	Val(2p)	1.17738	-0.06851
23	C	3	px	Ryd(3p)	0.00453	1.00792
24	C	3	py	Val(2p)	1.10869	-0.05230
25	C	3	py	Ryd(3p)	0.00377	0.79286
26	C	3	pz	Val(2p)	1.02987	-0.11312

FIGURE 2.9 The output result of optimized structure of the Lutidine with NAO (Natural Atomic Orbitals Occupancies).

The table succinctly illustrates the distribution of molecular charges using NPA charges. At this level of analysis, the C1 atom exhibits a net NPA charge of −0.59386 e. Notably, significant negative charges are observed on N5, C7, and C8. Additionally, a summary of the total populations in the natural minimal basis (NMB) and the natural Rydberg basis (NRB) [4] for the composite system is provided, calculated by aggregating the populations of individual atoms. Each atom is represented by an effective valence electron configuration referred to as the "Natural Electron Configuration" in natural populations. While atomic orbitals in a molecular context may exhibit noninteger occupancies, their effective configurations can be correlated with idealized atomic states in "promoted" configurations (Figure 2.11).

The table shown in Figure 2.12 provides an overview of the following data for each cycle: the number of low-occupancy Lewis (L) and high-occupancy (>0.1e)

Summary of Natural Population Analysis:

Atom	No	Natural Charge	Core	Valence	Rydberg	Total
			Natural Population			
C	1	-0.71533	1.99924	4.71160	0.00449	6.71533
C	2	0.00314	1.99894	3.98662	0.01130	5.99686
C	3	-0.28519	1.99891	4.27649	0.00979	6.28519
C	4	0.00422	1.99910	3.97755	0.01913	5.99578
N	5	-0.45723	1.99933	5.44732	0.01058	7.45723
C	6	0.20938	1.99901	3.77364	0.01797	5.79062
C	7	-0.72096	1.99923	4.71632	0.00541	6.72096
C	8	-0.27422	1.99885	4.26610	0.00928	6.27422
H	9	0.25137	0.00000	0.74749	0.00114	0.74863
H	10	0.25726	0.00000	0.74157	0.00117	0.74274
H	11	0.24862	0.00000	0.75025	0.00112	0.75138
H	12	0.24453	0.00000	0.75412	0.00135	0.75547
H	13	0.23503	0.00000	0.76341	0.00155	0.76497
H	14	0.25965	0.00000	0.73894	0.00141	0.74035
H	15	0.23755	0.00000	0.76144	0.00101	0.76245
H	16	0.25976	0.00000	0.73883	0.00141	0.74024
H	17	0.24241	0.00000	0.75627	0.00132	0.75759
* Total *		0.00000	15.99260	41.90795	0.09945	58.00000

	Natural Population
Core	15.99260 (99.9537% of 16)
Valence	41.90795 (99.7808% of 42)
Natural Minimal Basis	57.90055 (99.8285% of 58)
Natural Rydberg Basis	0.09945 (0.1715% of 58)

Atom	No	Natural Electron Configuration
C	1	[core]2S(1.09)2p(3.62)
C	2	[core]2S(0.88)2p(3.10)3p(0.01)
C	3	[core]2S(0.96)2p(3.32)3p(0.01)
C	4	[core]2S(0.93)2p(3.05)3p(0.02)
N	5	[core]2S(1.38)2p(4.07)3p(0.01)

FIGURE 2.10 The figure shows the summary of the natural population analysis.

non-Lewis (NL) orbitals; the maximum deviation of any formal bond order from an estimated value (NAO Wiberg bond index) for the structure; the occupancy threshold believed indicative of a "good" pair in the NBO search; the total populations of both Lewis and non-Lewis NBOs; the count of core (CR), 2-center bond (BD), 3-center bond (3C), and lone pair (LP) NBOs within the natural Lewis structure; and the total populations of both Lewis and non-Lewis NBOs. If the latter is higher than 0.1, further NBO searches are started.

The accuracy of the natural Lewis structure representation as a percentage of the total electron density (around 99.4% in this case) is shown in the next figure. The chart also illustrates how valence non-Lewis orbitals, as opposed to extra-valence orbitals (such Rydberg NBOs, which emerge when using a nonminimal basis set), significantly contribute to small deviations from a strictly localized Lewis structure model. The list of NBOs is given in the next section together with information on their form and occupancy for the whole set of NBOs that cover the input AO space (Figure 2.13).

```
NATURAL BOND ORBITAL ANALYSIS:

                       Occupancies       Lewis Structure    Low   High
          Occ.    -------------------  -----------------   occ    occ
 Cycle   Thresh.   Lewis   Non-Lewis    CR  BD  3C  LP     (L)   (NL)   Dev
=============================================================================
  1(1)    1.90   54.84241   3.15759     8  17   0   4      3     3     0.44
  2(2)    1.90   54.84241   3.15759     8  17   0   4      3     3     0.44
  3(1)    1.80   54.84241   3.15759     8  17   0   4      3     3     0.44
  4(2)    1.80   54.84241   3.15759     8  17   0   4      3     3     0.44
  5(1)    1.70   55.40051   2.59949     8  18   0   3      2     3     0.44
  6(2)    1.70   55.40051   2.59949     8  18   0   3      2     3     0.44
  7(1)    1.60   56.61604   1.38396     8  20   0   1      0     3     0.42
  8(2)    1.60   56.61345   1.38655     8  20   0   1      0     3     0.44
  9(3)    1.60   56.61604   1.38396     8  20   0   1      0     3     0.42
 10(1)    1.50   56.61604   1.38396     8  20   0   1      0     3     0.42
 11(2)    1.50   56.61345   1.38655     8  20   0   1      0     3     0.44
 12(3)    1.50   56.61604   1.38396     8  20   0   1      0     3     0.42
 13(1)    1.60   56.61604   1.38396     8  20   0   1      0     3     0.42
-----------------------------------------------------------------------------

Structure accepted: RESONANCE keyword permits strongly delocalized structure

WARNING:  1 low occupancy (<1.9990e) core orbital  found on  C   2
          1 low occupancy (<1.9990e) core orbital  found on  C   3
          1 low occupancy (<1.9990e) core orbital  found on  C   8

--------------------------------------------------------
  Core                      15.99261 ( 99.954% of  16)
  Valence Lewis             40.62343 ( 96.722% of  42)
 ==================         ============================
  Total Lewis               56.61604 ( 97.614% of  58)
 -------------------------------------------------------
  Valence non-Lewis          1.30576 (  2.251% of  58)
  Rydberg non-Lewis          0.07820 (  0.135% of  58)
 ==================         ============================
  Total non-Lewis            1.38396 (  2.386% of  58)
--------------------------------------------------------
```

FIGURE 2.11 The image shows the natural bond orbital analysis.

```
WARNING:  1 low occupancy (<1.9990e) core orbital  found on  C   8

--------------------------------------------------------
  Core                      15.99321 ( 99.958% of  16)
  Valence Lewis             40.58775 ( 96.637% of  42)
 ==================         ============================
  Total Lewis               56.58096 ( 97.553% of  58)
 -------------------------------------------------------
  Valence non-Lewis          1.31229 (  2.263% of  58)
  Rydberg non-Lewis          0.10675 (  0.184% of  58)
 ==================         ============================
  Total non-Lewis            1.41904 (  2.447% of  58)
--------------------------------------------------------
```

FIGURE 2.12 Overview of the critical parameters.

The next section provides an overview of the NBO basis's second-order perturbative estimations of "donor–acceptor" (bonded/anti-bonded) interactions.

In this study, all potential interactions between "empty" (acceptor) non-Lewis NBOs and "full" (donor) Lewis-type NBOs are investigated, and their energy relevance is estimated using 2nd-order perturbation theory (Figure 2.14). As a result of causing occupancy to be lost from the localized NBOs of the idealized Lewis

```
     (Occupancy)   Bond orbital/ Coefficients/ Hybrids
 ---------------------------------------------------------------------------------
   1. (1.98361) BD ( 1) C   1 - C   2
                ( 49.05%)   0.7004* C   1 s( 30.29%)p 2.30( 69.66%)d 0.00(  0.05%)
                                           0.0003  0.5501  0.0183  0.0010  0.0000
                                          -0.7332  0.0114  0.0070 -0.0008  0.3983
                                          -0.0063 -0.0049  0.0006 -0.0155  0.0010
                                          -0.0016 -0.0001 -0.0158  0.0000  0.0000
                                           0.0102 -0.0101
                ( 50.95%)   0.7138* C   2 s( 31.60%)p 2.16( 68.37%)d 0.00(  0.03%)
                                          -0.0001  0.5621  0.0032 -0.0045  0.0003
                                           0.7253  0.0067 -0.0052 -0.0023 -0.3969
                                          -0.0036  0.0025  0.0014  0.0028  0.0008
                                          -0.0003  0.0002 -0.0121  0.0007 -0.0004
                                           0.0075 -0.0096
   2. (1.98401) BD ( 1) C   1 - H   9
                ( 60.52%)   0.7779* C   1 s( 23.37%)p 3.28( 76.57%)d 0.00(  0.06%)
                                          -0.0001  0.4834 -0.0047 -0.0004  0.0000
                                           0.5979 -0.0058  0.0012  0.0003  0.4151
                                           0.0161  0.0007 -0.0012 -0.4854 -0.0031
                                          -0.0020 -0.0003  0.0147 -0.0148 -0.0117
                                           0.0034 -0.0009
                ( 39.48%)   0.6283* H   9 s( 99.97%)p 0.00(  0.03%)
                                           0.9998  0.0007  0.0005  0.0000 -0.0127
                                          -0.0078  0.0110
   3. (1.97316) BD ( 1) C   1 - H  10
                ( 60.73%)   0.7793* C   1 s( 22.82%)p 3.38( 77.12%)d 0.00(  0.06%)
                                          -0.0001  0.4777 -0.0074  0.0006  0.0002
                                           0.3132 -0.0113 -0.0013  0.0010 -0.0501
                                           0.0087  0.0011 -0.0008  0.8187  0.0065
                                           0.0029  0.0004  0.0000  0.0128 -0.0012
                                           0.0020  0.0209
                ( 39.27%)   0.6266* H  10 s( 99.97%)p 0.00(  0.03%)
                                           0.9998  0.0010  0.0004  0.0001 -0.0060
                                           0.0011 -0.0167
   4. (1.98751) BD ( 1) C   1 - H  11
                ( 60.41%)   0.7773* C   1 s( 23.54%)p 3.25( 76.41%)d 0.00(  0.06%)
                                           0.0001 -0.4851  0.0045  0.0006  0.0000
                                           0.0732  0.0172 -0.0003 -0.0013  0.8157
                                           0.0049  0.0022 -0.0006  0.3049  0.0019
```

FIGURE 2.13 Overviews the information on their form and occupancy for the whole set of NBOs calculations.

structure into the empty non-Lewis orbitals, these interactions are known as "delocalization" corrections to the zeroth-order natural Lewis structure. This leads to departures from the idealized Lewis structure description.

In summary, the list of the main NBOs that includes each one's occupancy, orbital energy, and qualitative pattern of delocalization interactions is shown in Figure 2.15.

2.3 HOMO-LUMO ANALYSIS

In molecular orbital (MO) theory, a mathematical equation has been used to describe the electrons delocalization between two specific atoms. The combination of atomic orbitals to make MOs is called the linear combination of atomic orbitals (LCAO). An atomic orbital corresponds to a specific atom and occupies a particular region of space, whereas a MO is related to the entire molecule and covers a larger area.

Two MOs are crucial for each molecule. The first is known as the highest occupied molecular orbital (HOMO), or the highest energy MO with any electrons present. The second is the lowest energy location where an electron may be placed or excited, known

```
Second Order Perturbation Theory Analysis of Fock Matrix in NBO Basis

    Threshold for printing:   0.50 kcal/mol
                                                                    E(2)  E(j)-E(i) F(i,j)
        Donor NBO (i)                     Acceptor NBO (j)        kcal/mol   a.u.    a.u.
===========================================================================================

within unit  1
  1. BD (   1) C   1 - C   2       / 64. RY*(   1) C   3                1.49    1.85    0.047
  1. BD (   1) C   1 - C   2       /149. RY*(   1) C   8                1.41    1.78    0.045
  1. BD (   1) C   1 - C   2       /222. BD*(   1) C   1 - H  10        0.55    1.00    0.021
  1. BD (   1) C   1 - C   2       /224. BD*(   1) C   2 - C   3        1.81    1.22    0.042
  1. BD (   1) C   1 - C   2       /225. BD*(   1) C   2 - C   8        1.87    1.22    0.043
  1. BD (   1) C   1 - C   2       /227. BD*(   1) C   3 - C   4        2.12    1.21    0.045
  1. BD (   1) C   1 - C   2       /235. BD*(   1) C   6 - C   8        2.22    1.22    0.047
  2. BD (   1) C   1 - H   9       /225. BD*(   1) C   2 - C   8        3.49    1.09    0.055
  2. BD (   1) C   1 - H   9       /226. BD*(   2) C   2 - C   8        1.65    0.54    0.029
  3. BD (   1) C   1 - H  10       / 49. RY*(   3) C   2                0.83    1.36    0.030
  3. BD (   1) C   1 - H  10       /225. BD*(   1) C   2 - C   8        0.87    1.09    0.027
  3. BD (   1) C   1 - H  10       /226. BD*(   2) C   2 - C   8        4.80    0.54    0.049
  4. BD (   1) C   1 - H  11       /224. BD*(   1) C   2 - C   3        4.14    1.09    0.060
  4. BD (   1) C   1 - H  11       /226. BD*(   2) C   2 - C   8        0.54    0.54    0.017
  5. BD (   1) C   2 - C   3       / 30. RY*(   1) C   1                0.98    1.29    0.032
  5. BD (   1) C   2 - C   3       / 81. RY*(   1) C   4                1.74    1.74    0.049
  5. BD (   1) C   2 - C   3       / 84. RY*(   4) C   4                1.05    2.29    0.044
  5. BD (   1) C   2 - C   3       /149. RY*(   1) C   8                1.23    1.85    0.043
  5. BD (   1) C   2 - C   3       /220. BD*(   1) C   1 - C   2        1.54    1.12    0.037
  5. BD (   1) C   2 - C   3       /223. BD*(   1) C   1 - H  11        0.68    1.10    0.024
  5. BD (   1) C   2 - C   3       /225. BD*(   1) C   2 - C   8        3.17    1.28    0.057
  5. BD (   1) C   2 - C   3       /227. BD*(   1) C   3 - C   4        2.77    1.27    0.053
  5. BD (   1) C   2 - C   3       /229. BD*(   1) C   3 - H  12        1.11    1.14    0.032
  5. BD (   1) C   2 - C   3       /231. BD*(   1) C   4 - H  13        2.04    1.13    0.043
  5. BD (   1) C   2 - C   3       /239. BD*(   1) C   8 - H  17        2.87    1.13    0.051
  6. BD (   1) C   2 - C   8       / 30. RY*(   1) C   1                0.88    1.29    0.030
  6. BD (   1) C   2 - C   8       / 64. RY*(   1) C   3                1.29    1.91    0.045
  6. BD (   1) C   2 - C   8       /115. RY*(   1) C   6                1.81    1.75    0.050
  6. BD (   1) C   2 - C   8       /117. RY*(   3) C   6                0.98    2.03    0.040
  6. BD (   1) C   2 - C   8       /220. BD*(   1) C   1 - C   2        1.54    1.12    0.037
  6. BD (   1) C   2 - C   8       /221. BD*(   1) C   1 - H   9        0.56    1.10    0.022
```

FIGURE 2.14 Overview of the NBO basis' second-order perturbative estimations of "donor-acceptor" (bonded/anti-bonded) interactions.

as the lowest unoccupied molecular orbital (LUMO). Since these two orbitals have the lowest energy separation compared to every orbital in a molecule, the energy difference between them is the most significant one to take into account. An electronic excitation in a molecule requires a minimum amount of energy known as the HOMO-LUMO energy gap, which is the difference in energy between the HOMO and the LUMO.

In chemical reactions, the LUMO and the HOMO are most significant. An electron must move from the HOMO to the LUMO in order for a molecule to undergo a reaction. Because it requires less energy for molecules with shorter HOMO-LUMO gaps to produce this phenomenon, they are more reactive.

2.3.1 Calculation Steps

Here, the calculation was performed using phenol molecule as an example. Initially, the molecule was optimized using 6-31G/B3LYP level of DFT as shown in Figures 2.16–2.18.

For single-point calculation, again open the calculation page of optimized file. In the Gaussian calculation window, select energy in the job type section.

Method, along with section and other options, remains the same in the single-point calculation.

After successfully a single-point calculation is done, open the log file of the single-point file. Click the tool and select MOs. This is shown in Figure 2.19.

```
                                                             Principal Delocalizations
           NBO                        Occupancy   Energy    (geminal,vicinal,remote)
=======================================================================================
Molecular unit  1  (C7H9N)
    1. BD (   1) C   1 - C   2          1.98361    -0.64102  235(v),227(v),225(g),224(g)
                                                   64(v),149(v),222(g)
    2. BD (   1) C   1 - H   9          1.98401    -0.51100  225(v),226(v)
    3. BD (   1) C   1 - H  10          1.97316    -0.50814  226(v),225(v),49(v)
    4. BD (   1) C   1 - H  11          1.98761    -0.51183  224(v),226(v)
    5. BD (   1) C   2 - C   3          1.97373    -0.70239  225(g),239(v),227(g),231(v)
                                                   81(v),220(g),149(v),229(g)
                                                   84(v),30(v),223(v)
    6. BD (   1) C   2 - C   8          1.97344    -0.70238  235(g),234(v),224(g),229(v)
                                                   115(v),220(g),64(v),239(g)
                                                   117(v),30(v),221(v)
    7. BD (   2) C   2 - C   8          1.64187    -0.25583  233(v),228(v),222(v),66(v)
                                                   221(v),121(v),31(v)
    8. BD (   1) C   3 - C   4          1.98307    -0.70910  220(v),224(g),230(g),48(v)
                                                   99(v),229(g),47(v),231(g)
    9. BD (   2) C   3 - C   4          1.65250    -0.25826  226(v),233(v),100(v),66(g)
                                                   54(v)
   10. BD (   1) C   3 - H  12          1.97944    -0.52394  230(v),225(v),47(v),82(v)
                                                   224(g),227(g)
   11. BD (   1) C   4 - N   5          1.98450    -0.81860  116(v),234(v),227(g),229(v)
                                                   65(v),232(g)
   12. BD (   1) C   4 - H  13          1.98226    -0.51854  232(v),224(v),64(v),98(v)
   13. BD (   1) N   5 - C   6          1.98336    -0.81857  82(v),231(v),235(g),239(v)
                                                   230(g),234(g),237(v),132(v)
                                                   150(v)
   14. BD (   2) N   5 - C   6          1.70415    -0.29206  228(v),226(v),83(v),236(v)
                                                   238(v),133(v),153(v)
   15. BD (   1) C   6 - C   7          1.98280    -0.62833  230(v),225(v),235(g),98(v)
                                                   232(g),149(v),236(g),238(g)
   16. BD (   1) C   6 - C   8          1.97928    -0.70477  220(v),225(g),232(g),48(v)
                                                   234(g),47(v),239(g),132(v)
                                                   98(v)
   17. BD (   1) C   7 - H  14          1.97686    -0.49593  233(v),235(v)
   18. BD (   1) C   7 - H  15          1.98880    -0.50312  232(v),116(v)
   19. BD (   1) C   7 - H  16          1.97701    -0.49592  233(v),235(v)
```

FIGURE 2.15 Provides the result summary of NBO calculations.

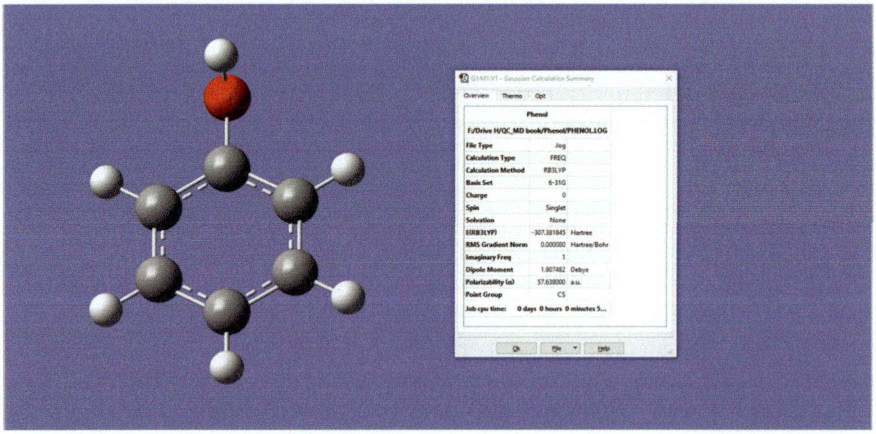

FIGURE 2.16 The optimized Phenol molecule with respective calculation result summary.

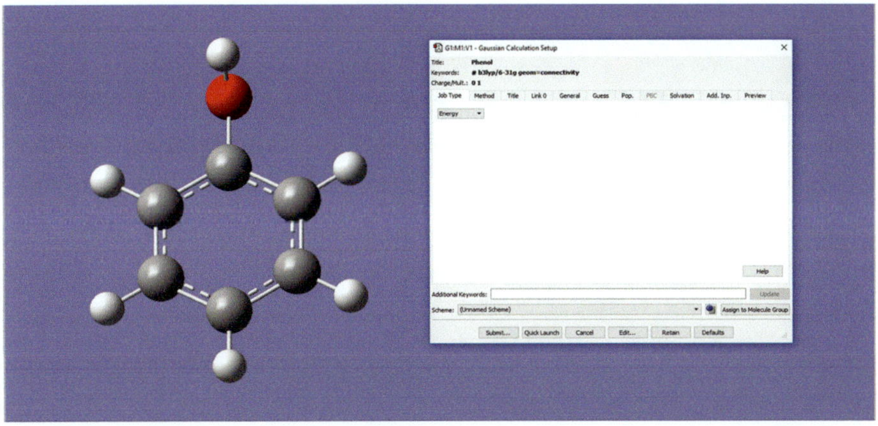

FIGURE 2.17 The optimized phenol molecule with respective Gaussian calculation setup.

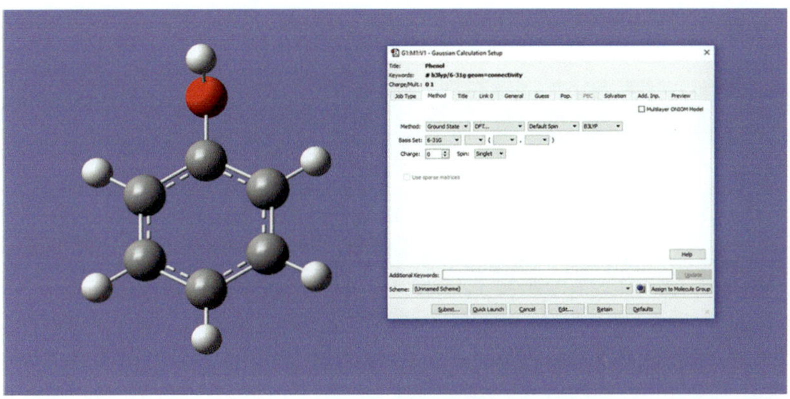

FIGURE 2.18 The optimized phenol molecule with the respective Gaussian calculation setup with method tab details.

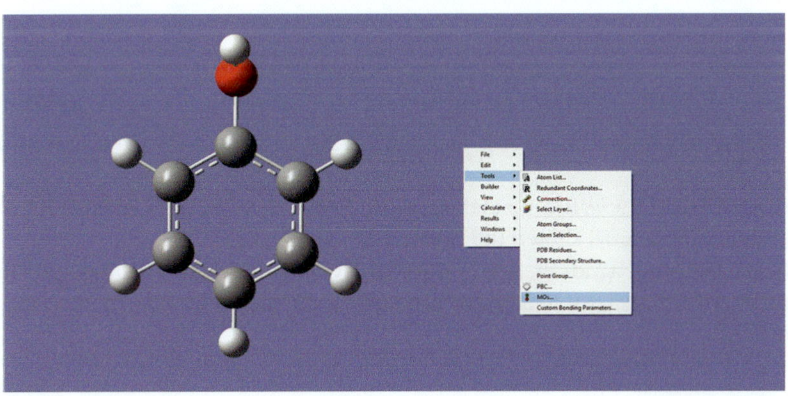

FIGURE 2.19 The optimized phenol molecule with respective steps to selects MOs.

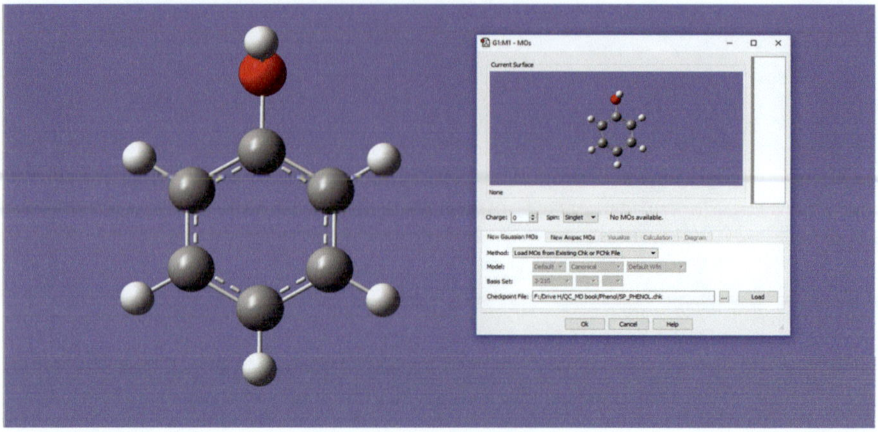

FIGURE 2.20 The optimized phenol molecule with respective Gaussian calculation setup to create new Gaussian MOs and load .chk file.

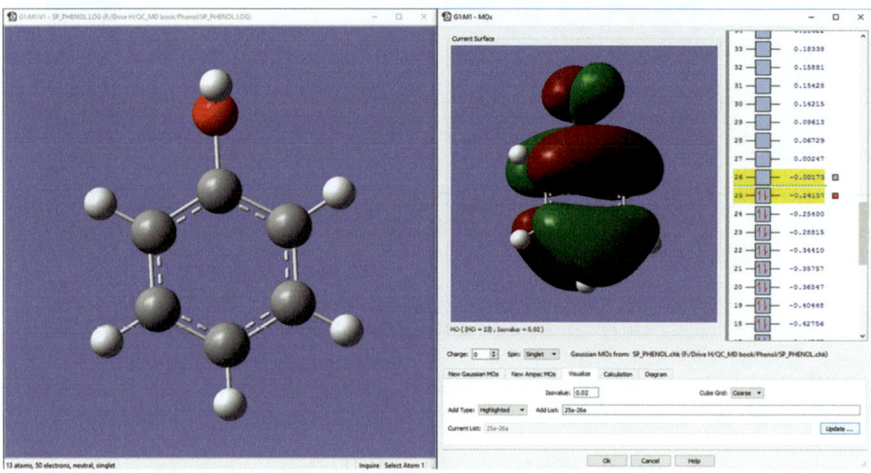

FIGURE 2.21 The HOMO-LUMO clouds for the phenol molecule.

In the MOs window, as shown in Figure 2.20, click on New Gaussian MOs and load the *.chk* file.

After loading the *.chk* file, go to the *visualize* and click *update*. The HOMO-LUMO clouds should appear as shown in Figure 2.21.

After successfull loading, a HOMO and LUMO diagram will appear. The values of the HOMO-LUMO diagram are shown on the right side of the images in Figure 2.22.

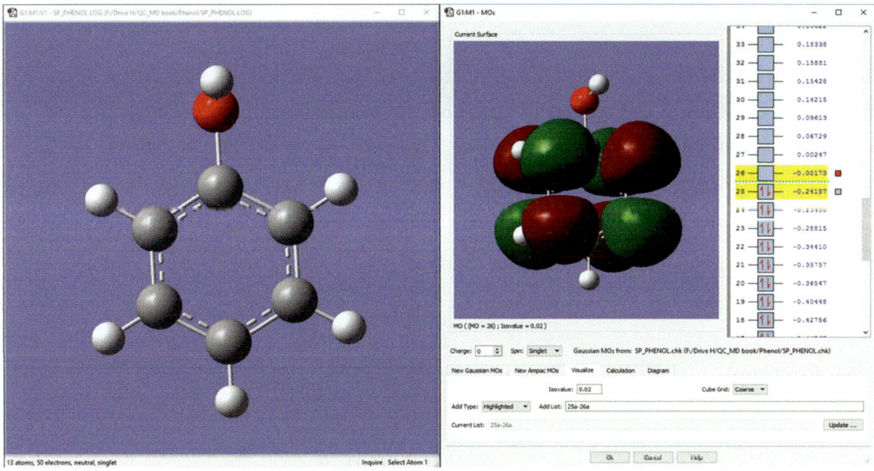

FIGURE 2.22 The HOMO-LUMO values for the phenol molecule.

2.4 GLOBAL SCALAR PROPERTIES

HOMO signifies the capacity to donate an electron, whereas LUMO signifies the capability to receive an electron, as described in Section 2.3. HOMO and LUMO are widely recognized as the principal frontier orbitals that contribute to chemical stability. As the highest energy orbital is capable of containing electrons, HOMO energy is strongly associated with reactivity to electrophilic attack; conversely, LUMO energy is closely associated with reactivity to nucleophilic attack, as it is the lowest energy orbital capable of accepting electrons. Gaining valuable theoretical understanding of selectivity and reactivity parameters in the chemical realm can be accomplished through the determination of globally recognized qualitative chemical concepts, including the global hardness (η), global softness (S), electrophilicity index (ω), electronegativity (χ), and chemical potential (μ). It is possible to compute these properties by employing the HOMO and LUMO energy values.

The LUMO's energy is intrinsically linked to the electron affinity (EA) and serves as an indicator of the molecule's vulnerability to nucleophilic attack. The term EA denotes the specific capacity of a ligand to accept one electron from a donor. The HOMO's energy is intrinsically linked to the ionization potential (IP) and serves as a measure of the molecules' vulnerability to electrophilic attack. The operational definition of LUMO and HOMO in the context of DFT can be written as

$$\mathrm{EA} = -\xi_{\mathrm{LUMO}} \tag{2.2}$$

$$\mathrm{IP} = -\xi_{\mathrm{HOMO}} \tag{2.3}$$

ξ_{HOMO} and ξ_{LUMO} correspond to Kohn–Sham eigenvalues associated to the frontier MOs. The definition of the global scalar properties is as follows:

TABLE 2.1
Global Scalar Properties and Its Definitions with Values Associated to HOMO and LUMO Values of Phenol Molecule Studied in Section 2.3 at 6–31G Functional and B3LYP Level of Theory

Properties	Expression	Definition	Values
Electronegativity (χ)	$\left(\frac{IP+EA}{2}\right)$	$-\left(\frac{\xi_{HOMO}+\xi_{LUMO}}{2}\right)$	−0.12165
Chemical potential (μ)	$-\left(\frac{IP+EA}{2}\right)$	$\left(\frac{\xi_{HOMO}+\xi_{LUMO}}{2}\right)$	0.12165
Chemical or global hardness (η)	$\left(\frac{IP-EA}{2}\right)$	$\left(\frac{-\xi_{HOMO}+\xi_{LUMO}}{2}\right)$	0.11992
Chemical or global softness (σ)	$\left(\frac{2}{IP-EA}\right)$	$\left(\frac{2}{-\xi_{HOMO}+\xi_{LUMO}}\right)$	8.33889
Electrophilicity index (ω)	$\left(\frac{\frac{(IP+EA^{2}}{2)}}{(IP-EA)}\right)$	$\left(\frac{\frac{(\xi_{HOMO}+\xi_{LUMO}{}^{2}}{2)}}{(-\xi_{HOMO}+\xi_{LUMO})}\right)$	0.24503

2.4.1 Chemical Potential (μ)

It is a measure of an electron cloud's escape tendency and is known as the first derivative of total energy with regard to the number of electrons in a molecule [5].

2.4.2 Global Hardness and Softness

Molecular hardness, denoted by the symbol η, serves as a qualitative indicator of the degree to which the molecule is polarizable. Hard molecules are characterized by a large HOMO–LUMO energy gap, signifying their exceptional stability and resistance to alterations in electron number and distribution. A small HOMO–LUMO energy gap signifies a high degree of polarizability in soft molecules (global or chemical softness), which can be attributed to their need of low excitation energy [6].

Electrophilicity Index: Electrophilicity index (ω) quantifies a molecule's ability to accept electrons by measuring the energy reduction resulting from electron transfer between donor and acceptor [7].

2.4.3 Electronegativity

Electronegativity (χ) is a measure of the ability of a molecule to draw electrons toward itself, determined by the mean value of the HOMO and LUMO energy levels. Electronegativity (χ) and hardness are commonly utilized to predict the chemical behavior of compounds [8]. In Section 2.3, the global scalar characteristics of the phenol molecule were calculated. The HOMO and LUMO values were found to be −0.24157 and −0.00173 Hartrees, respectively. We will be using these values to predict the global scalar properties defined here (Table 2.1).

ACKNOWLEDGMENTS

We are grateful to Gaussian and Gaussview Inc. for allowing us to use their soft ware's logo and images in this chapter (see references [9,10]).

REFERENCES

1. Weinhold, F., C. Landis, and E. Glendening, What is NBO analysis and how is it useful? *International Reviews in Physical Chemistry*, 2016. 35(3): p. 399–440.
2. Kolandaivel, P. and V. Nirmala, Study of proper and improper hydrogen bonding using Bader's atoms in molecules (AIM) theory and NBO analysis. *Journal of Molecular Structure*, 2004. 694(1–3): p. 33–38.
3. Szafran, M., A. Komasa, and E. Bartoszak-Adamska, Crystal and molecular structure of 4-carboxypiperidinium chloride (4-piperidinecarboxylic acid hydrochloride). *Journal of Molecular Structure*, 2007. 827(1–3): p. 101–107.
4. Reed, A.E., R.B. Weinstock, and F. Weinhold, Natural population analysis. *The Journal of Chemical Physics*, 1985. 83(2): p. 735–746.
5. Gázquez, J.L., Perspectives on the density functional theory of chemical reactivity. *Journal of the Mexican Chemical Society*, 2008. 52(1): p. 3–10.
6. Omar, S., M. Shkir, and M.A. Khan, et al., A comprehensive study on molecular geometry, optical, HOMO-LUMO, and nonlinear properties of 1, 3-diphenyl-2-propen-1-ones chalcone and its derivatives for optoelectronic applications: a computational approach. *Optik*, 2020. 204: p. 164172.
7. Pal, R. and P.K. Chattaraj, Electrophilicity index revisited. *Journal of Computational Chemistry*, 2023. 44(3): p. 278–297.
8. Pearson, R.G., Chemical hardness and density functional theory. *Journal of Chemical Sciences*, 2005. 117: p. 369–377.
9. GaussView, Version 6, Dennington, Roy; Keith, Todd A.; Millam, John M. Semichem Inc., Shawnee Mission, KS, 2016.
10. Gaussian 09, Revision D.01, Frisch, M. J.; Trucks, G. W.; Schlegel, H. B.; Scuseria, G. E.; Robb, M. A.; Cheeseman, J. R.; Scalmani, G.; Barone, V.; Petersson, G. A.; Nakatsuji, H.; Li, X.; Caricato, M.; Marenich, A. V.; Bloino, J.; Janesko, B. G.; Gomperts, R.; Mennucci, B.; Hratchian, H. P.; Ortiz, J. V.; Izmaylov, A. F.; Sonnenberg, J. L.; Williams-Young, D.; Ding, F.; Lipparini, F.; Egidi, F.; Goings, J.; Peng, B.; Petrone, A.; Henderson, T.; Ranasinghe, D.; Zakrzewski, V. G.; Gao, J.; Rega, N.; Zheng, G.; Liang, W.; Hada, M.; Ehara, M.; Toyota, K.; Fukuda, R.; Hasegawa, J.; Ishida, M.; Nakajima, T.; Honda, Y.; Kitao, O.; Nakai, H.; Vreven, T.; Throssell, K.; Montgomery, J. A., Jr.; Peralta, J. E.; Ogliaro, F.; Bearpark, M. J.; Heyd, J. J.; Brothers, E. N.; Kudin, K. N.; Staroverov, V. N.; Keith, T. A.; Kobayashi, R.; Normand, J.; Raghavachari, K.; Rendell, A. P.; Burant, J. C.; Iyengar, S. S.; Tomasi, J.; Cossi, M.; Millam, J. M.; Klene, M.; Adamo, C.; Cammi, R.; Ochterski, J. W.; Martin, R. L.; Morokuma, K.; Farkas, O.; Foresman, J. B.; Fox, D. J. Gaussian, Inc., Wallingford CT, 2016.

3 Theory of Hydrogen Bonding

Atoms-in-Molecules, and Non-Covalent Interaction Analysis

3.1 THEORY OF HYDROGEN BONDING

H-bond concept was first coined by Pimental and McClellan in 1960 considered one of the most precursor ideas. The first intramolecular H-bonding was suggested by Oddo and Puxeddu in 1906 to explain the properties of some o-hydroxyazo derivatives of eugenol and after that, in 1913, G. Gilli and P. Gilli [1] proposed a similar idea about the intermolecular link, named as inner Komplexsalzbindung, which is formed between the hydroxyl and carbonyl functions of 1-hydroxy-anthraquinone. The details of H-bonding were given by Latimer and Rodebush [2] in 1920, which showed that the hydrogen bond was responsible for the dielectric constant and boiling point .

The modern definition of hydrogen bonding recognized by IUPAC is given as

> The hydrogen bond is an attractive interaction between a hydrogen atom from a molecule or a molecular fragment X–H in which X is more electronegative than H, and an atom or a group of atoms in the same or another molecule, in which there is evidence of bond formation [3]

A hydrogen bond is a type of intermolecular contact that occurs between a hydrogen atom with a partial positive charge and an area of high electron density [4]. Hydrogen can be formed in the form of D–H---A, where D is the hydrogen bond donor and A is the hydrogen bond acceptor. One example of H-bonding formation is water molecules, as shown in Figure 3.1. Water molecules form by the bonding of the extremely electronegative oxygen atom with a hydrogen atom. The shared pairs of electrons are strongly attracted to the oxygen atoms, resulting in a negative charge at one end of the molecule. Alternatively, the hydrogen atoms have a positive charge.

Hydrogen bond has the ability of forming one chemical bond with another compound; however, hydrogen is formally two valent, and the additional bond is called a hydrogen bond. Hydrogen bonds can be classified into two categories:

1. Hydrogen bonds that connect atoms of higher electronegativity
2. Hydrogen bonds that connect atoms of lower electronegativity

DOI: 10.1201/9781032694733-3

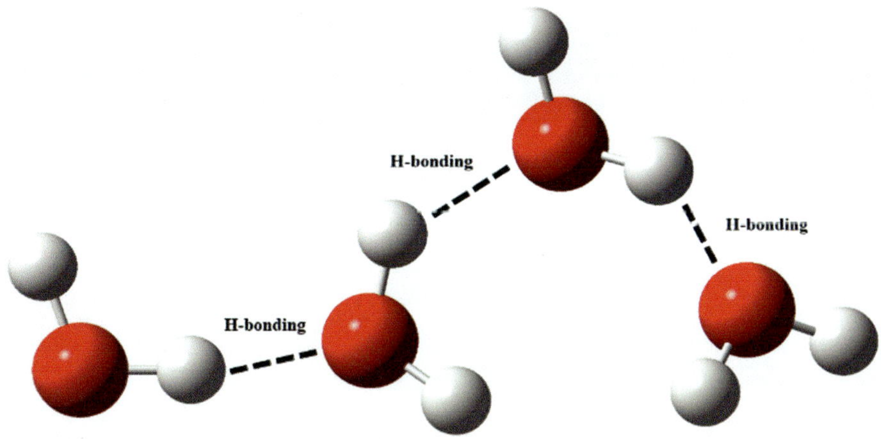

FIGURE 3.1 Hydrogen bonding between water molecules, showing oxygen and hydrogen atoms with partial charges, and hydrogen bond interactions between molecules.

The elements in the periodic table that have electronegativity values higher than hydrogen include carbon (C), nitrogen (N), oxygen (O), fluorine (F), phosphorus (P), sulfur (S), chlorine (Cl), selenium (Se), bromine (Br), and iodine (I). Hydrogen bonds are formed between all of these components. Additionally, "π" hydrogen bonds occur when a partially positive hydrogen interacts with the electrons in a double or triple bond. These bonds have also been the focus of some research.

The first theoretical models of hydrogen bonding initially focused solely on electrostatics, assuming that molecules involved in hydrogen bonding consisted of a combination of positive and negative charges. According to Coulson (1959) [5], the formation of hydrogen bonds involves four distinct concepts: The interaction between molecules can be attributed to four main factors: (a) electrostatic forces resulting from the interaction of undistorted charge clouds, (b) delocalization effects caused by significant deformations of these charge clouds, leading to partial covalent character in the formed hydrogen bond, (c) repulsive force caused by the Pauli exclusion principle due to the overlapping of closed shell charge clouds, and (d) London dispersion force resulting from a small displacement of electrons [6].

In summary, quantum chemical methods have been so far extremely successful and convincing in giving a detailed characterization of the structural, energetic, and electron density features of both weak and strong H-bond forms. Theoretical methods have not been equally successful in solving the H-bond puzzle, that is, explaining the bond formation through the same bond acceptor–donor couple can display different interaction energies [7].

3.2 INTRODUCTION TO MULTIWFN PACKAGE

Electronic wave function is the essential component of quantum chemistry. This can be easily investigated and analyzed through Multiwfn software, a free & open-source, highly efficient, very user-friendly, and flexible one. The software can run easily on

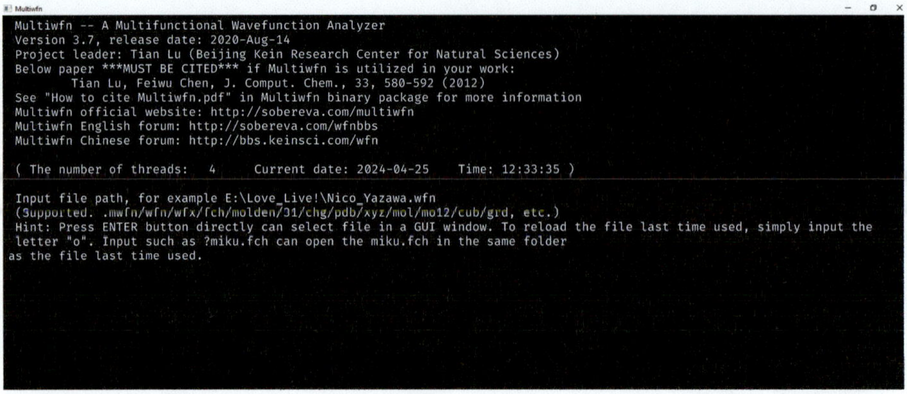

FIGURE 3.2 Screenshot of the Multiwfn main window displaying the graphical user interface and file loading options.

Windows and Linux environments. It is best used for visualization purpose that can obtain good visualization of the molecules. The screenshot image of multiwfn software is shown in Figure 3.2.

Some of the important features of multiwfn are [8]

1. Multiwfn is built as an interactive program that provides clear instructions to users at each step, ensuring that they know what to do next.
2. It prevents the printing of confusing messages, making it easy for beginners to understand. Furthermore, the handbook contains more than 20 practical examples that would be very helpful for newcomers.
3. Plotting does not require the use of any third-party software. Muliwfn utilizes its own graphical plot software library, DISLIN, which is automatically invoked internally for visualizing findings. Users may interactively alter most of the plotting settings using an interface.
4. In multiwfn, the majority of the systems based on the command line mode and a Graphical User Interface (GUI) appears when required. The advantage of using Multiwfn is that it can run in silent and batch mode easily by redirection and shell scripts.
5. Memory requirement in the Multiwfn software is very low, and it gives high efficiency. The code of Multiwfn is well optimized. The majority of components are parallelized using OpenMP technology. Multiwfn is more efficient than a similar program for jobs that need a lot of time.

3.2.1 Real Space Function Supported by Multiwfn

Visualization of real-space function is one of the important features of Multiwfn. It is a convenient way to study electron structures. Multiwfn offers a diverse array of valuable real-space functions, which will be introduced in sequence. The functions covered from Sections "electron density" to "sign(λ_2)$\times q$" rely solely on electron density. Consequently, they can be derived not only through quantum chemistry calculations but also through experimental methods such as X-ray diffraction. There are more than 100 real space functions supported in the Multiwfn software, and some of them are listed below:

1. Electron density
2. Gradient norm of electron density
3. Laplacian of electron density
4. Wave function value and probability density of an orbital
5. Electron spin density
6. Hamiltonian kinetic energy density
7. Lagrangian kinetic energy density
8. Electron localization function (ELF) defined by Becke and the one defined by Tsirelson
9. Localized orbital locator (LOL) defined by Becke and the one defined by Tsirelson
10. Interaction region indicator (IRI) and IRI-π
11. δg function defined in independent gradient model (IGM) and IGM based on Hirshfeld partition (IGMH)
12. Local information entropy
13. Electrostatic potential (ESP), and that from nuclear/electronic/atomic charges
14. van der Waals potential
15. Reduced density gradient (RDG) with/without promolecular approximation
16. sign(λ_2)ρ (product of the sign of the second largest eigenvalue of electron density Hessian matrix and electron density) with/without promolecular approximation
17. Exchange-correlation density, correlation hole, and correlation factor
18. Average local ionization energy (ALIE) and local electron attachment energy (LEAE)
19. Source function
20. Electron delocalization range function EDR (r; d) and orbital overlap distance function $D(r)$

The display of the main function menu of the multiwfn is shown in Figures 3.2 and 3.3.

3.2.2 Input File Supported by Multiwfn

Multiwfn is a software that can analyze wave functions using files generated by various quantum chemistry programs such as Gaussian, ORCA, GAMESS-US, Molpro, NWChem, Dalton, xtb, PSI4, Molcas, Q-Chem, MRCC, deMon2k, Firefly, CFOUR, and Turbomole. Additionally, it can also analyze periodic systems by supporting .molden files exported by CP2K. It can support .wfn/.wfx, .fch, .molden, and .gms. Other files that are important in the multiwfn software are .cub, .grd, .pdb, .xyz, and .mol files.

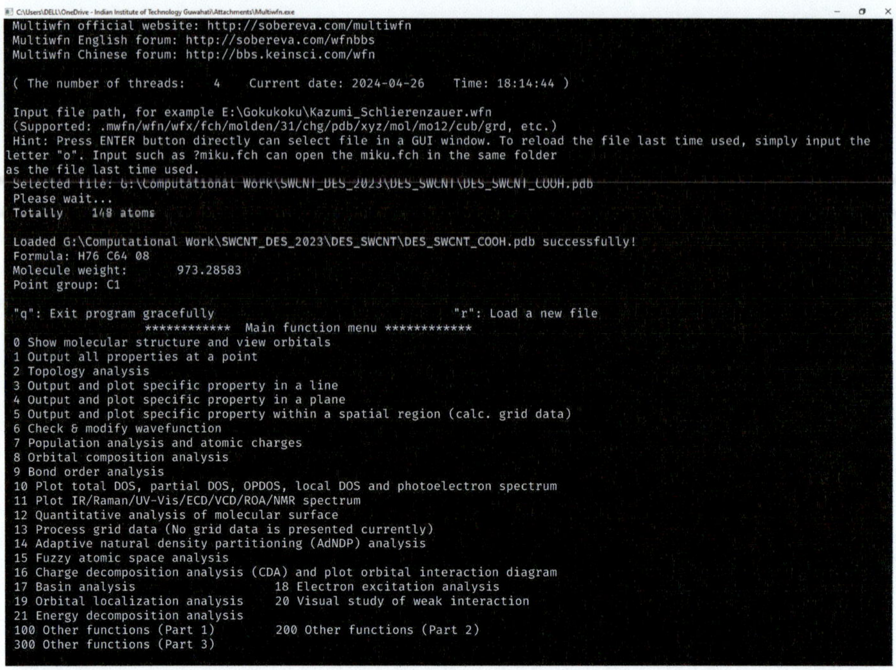

FIGURE 3.3 Screenshot of Multiwfn's main function window, showcasing various calculation options available for quantum chemical analysis.

3.2.3 Graphic Formats and Image Size

The graphic format supported by the Multiwfn software:

1. Postscript (ps)
2. Encapsulated postscript (eps)
3. Portable document format (pdf)
4. Windows metafile format (wmf)
5. Graphics interchange format (giff)
6. TIFF (tiff)
7. Portable network graphic (png)
8. Windows bitmap format (bmp)
9. Scalable vector graphics (svg)

The images were exported in a graphic format using Multiwfn, with the specific format being selected by the "graph format" parameter in the *settings.ini.* The "graph1Dsize" parameter in the *settings.ini* controls the dimensions of the height and weight of the curves image file. The function "graph2Dsize" is responsible for charting two-dimensional data, such as color-filled maps, contour maps, and relief maps. The function "graph3Dsize" is responsible for plotting three-dimensional data, such as iso-surface graphs and molecular structure graphs.

3.2.4 Details on Input Files and Wavefunction Types

Multiwfn supports various types of input files as detailed in the previous Section 3.2.2. Multiwfn is capable of handling a range of wavefunction types, such as single-determinant wavefunctions, open-shell wavefunctions, and post-HF wavefunctions expressed in natural orbital formalism. The package provides support for Gaussian functions with angular momentum up to h, applicable to both Cartesian and spherical harmonic forms. The software Multiwfn does not impose a defined upper limit on the quantity of atoms, basis functions, Gaussian type functions (GTFs), or orbitals that can be utilized. The limit is exclusively dictated by the available memory on your computer.

Multiwfn is capable of identifying the type of input file by analyzing its file extension. Selecting the appropriate type of input file for various functions is essential. Refer to the table below for guidance. The wavefunction represented by GTFs is adequate for performing Hirshfeld population calculations. Various file formats can be utilized as input, including .mwfn, .fch, .molden, .wfn, and .wfx. File formats such as .pdb, .xyz, and .mol do not include wavefunction information and are therefore not applicable for use. Conversely, the generation of grid data for the RDG function using the promolecular approximation necessitates only the coordinates of the atoms. Consequently, all supported file formats, with the exception of plain text files, are applicable for use. The types of information available in various file formats are presented in Table 3.1.

TABLE 3.1
Different Formats of File Used for Used by Multiwfn Software Package

	Contained Information Types				
File Format	**Basis Functions**	**Gaussian Type Functions**	**Atom Coordinates**	**Grid Data**	**Atomic Charges**
.fch/.fchk/.chk	✓	✓	✓	×	×
.mwfn, .molden, .gbw	✓	✓	✓	×	×
NBO .31~.40	×	✓	✓	×	×
.wfn and .wfx	×	✓	✓	×	×
.pdb, .xyz, .mol/.sdf, .cif, .mop, .gro	×	×	×	×	×
.chg and .pqr	×	×	✓	×	✓
.cub/.cube	×	×	✓	✓	×
.vti, .grd, .dx	×	×	×	✓	×

Let's discuss some important and more frequent file extensions used:

a. **Multiwfn wavefunction file (.mwfn):** This format has been defined and is supported starting from Multiwfn version 3.7. This format is optimal for the storage and exchange of wavefunctions. This file is structured to provide all essential information for wavefunction analysis in a strict, succinct, compact, and extensible format. The introduction and definition of this format have been meticulously detailed in paper published by Tian Lu and coworkers, ChemRxiv (2020), DOI: 10.26434/chemrxiv.11872524.
b. **AIM wavefunction file (.wfn):** The format was first introduced by Bader in the AIMPAC program and is currently supported by several widely used quantum chemistry software packages, such as Gaussian, ORCA, GAMESS-US/UK, Firefly, Q-Chem, and NWChem. The .wfn file contains atomic coordinates, elements, orbital energies, occupation numbers, and expansion coefficients of Cartesian Gaussian type functions (GTF). Here, AIM stands for ***atoms in molecules***.
c. **Gaussian formatted checkpoint file (.fch/.fchk):** The formchk utility within the Gaussian package facilitates the conversion of a Gaussian program checkpoint file (.chk) into a formatted checkpoint file (.fch/.fchk). The .fch and .fchk file formats are equivalent and do not exhibit any differences. The default extension produced by the Windows (Linux) version of formchk is "fch" ("fchk"). To configure Multiwfn for direct loading of .chk files, it is essential to specify "formchkpath" within the settings. Specify the actual path of the formchk executable file within the Gaussian package. Multiwfn will automatically invoke formchk to convert the .chk file to .fch/fchk file in this scenario. Upon successful conversion, the .fch/fchk file will be loaded and subsequently removed after the loading process is completed. The data present in .fch/.fchk files is more extensive compared to that in .wfn/.wfx files. Virtual orbital wavefunctions are recorded, providing basis function information for Multiwfn. To utilize .fch/.fchk files as carriers for post-HF wavefunctions, it is essential to review the Multiwfn package manual in detail.
d. **Molden input file (.molden, .molden.input, or molden.inp):** Currently, various quantum chemistry software packages such as Molpro, Molcas, ORCA, Q-Chem, CFour, Turbomole, PSI4, MRCC, and NWChem, along with the first-principles code CP2K, have the capability to produce input files for the Molden visualization program. This file type includes the definition of a basis set, atomic coordinates, and details regarding all occupied and virtual orbitals. This encompasses the expansion coefficients of basis functions, occupation numbers, spin, energy, and symmetry information. The Molden input file serves as a standard and comprehensive format for the exchange of wavefunction information. This type of file provides atomic coordinates, basis function information, and GTF information for use with the Multiwfn package.

3.3 PREPARATION OF .WFN, .CHK, AND .MOL FILE

3.3.1 .WFN FILE

If you require the **.wfn** file, you can produce it using Gaussian:

Step 1: Run a single-point energy calculation on an optimized geometry of a molecular system

Step 2: In a new calculation, specify the term **output=wfn** followed by **filename.wfn** after the coordinates. One important thing is to be considered please leave a line after the coordinates to provide the path for the **filename.wfn** as mentioned in Figure 3.4 below. One more thing like to be added if you want to search for generated .wfn file, the user should go to the directory path that you have provided while setting up the calculation (Figure 3.5).

3.3.2 .CHK FILE

To generate a .chk (checkpoint) file utilizing Gaussian, one must incorporate a line reading **"%chk=filename.chk"** at the commencement of the Gaussian input file, where "filename.chk" denotes the preferred designation for the checkpoint file. This directive instructs Gaussian to preserve the requisite data in a binary checkpoint file bearing that specified name throughout the computational process.

3.3.3 .MOL FILE

To create a .mol file using Gaussian, you need to perform a calculation on your molecule within Gaussian, then extract the final optimized geometry from the output log file (.log) and convert it to a .mol file using a molecular visualization program like GaussView or a third-party software.

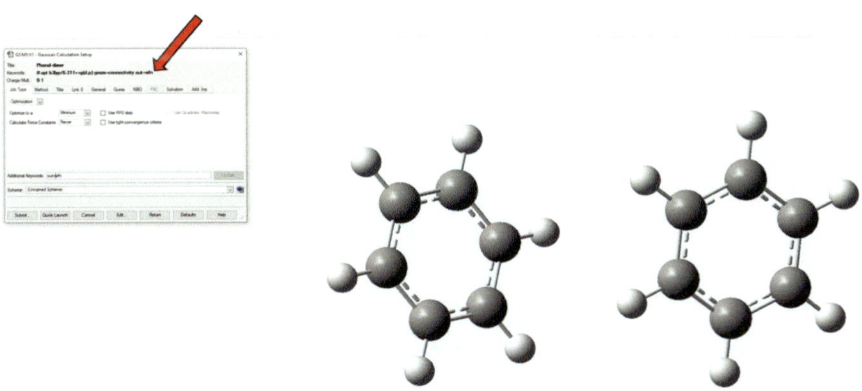

FIGURE 3.4 Screenshot of the setup for a single-point energy calculation on an optimized phenol dimer geometry for **.wfn** file generation.

```
%chk=D:\QC books\Chapter 3 \Phenol-dimer.chk
# opt b31yp/6-311++g(d,p) geom=connectivity out=wfn

Phenol-dimer

0 1
 C                  4.16812497    0.78922425   -2.47825406
 C                  3.07380939    0.15134200   -1.89458498
 C                  1.79452930    0.68723455   -2.04081133
 C                  1.60928114    1.86110491   -2.77074110
 C                  2.70359537    2.49897672   -3.35441643
 C                  3.98288146    1.96309380   -3.20818043
 H                  5.16281033    0.37256577   -2.36455977
 H                  3.21758219   -0.76142253   -1.32697207
 H                  0.94380053    0.19105047   -1.58685772
 H                  0.61459992    2.27777509   -2.88444239
 H                  2.55983742    3.41174141   -3.92203055
 H                  4.83360782    2.45927825   -3.66211829
 C                 -2.63405327    0.93002124    0.72260744
 C                 -3.85509333    0.78263450    1.38019942
 C                 -4.97598868    1.48310377    0.93516424
 C                 -4.87609653    2.33109802   -0.16756653
 C                 -3.65506413    2.47847942   -0.82516534
 C                 -2.53416459    1.77801756   -0.38011797
 H                 -1.76252572    0.38538411    1.06864575
 H                 -3.93304105    0.12333255    2.23765883
 H                 -5.92529902    1.36832856    1.44666320
 H                 -5.74763604    2.87572594   -0.51358700
 H                 -3.57712292    3.13778293   -1.68262584
 H                 -1.58486502    1.89283043   -0.89164351

 1 2 1.5  6 1.5  7 1.0
 2 3 1.5  8 1.0
 3 4 1.5  9 1.0
 4 5 1.5  10 1.0
 5 6 1.5  11 1.0
 6 12 1.0
 7
 8
 9
 10
 11
 12
 13 14 1.5  18 1.5  19 1.0
 14 15 1.5  20 1.0
 15 16 1.5  21 1.0
 16 17 1.5  22 1.0
 17 18 1.5  23 1.0
 18 24 1.0
 19
 20
 21
 22
 23
 24

D:\QC books\Chapter 3\Phenol-dimer.wfn
```

FIGURE 3.5 Example of a Gaussian input file used for generating a **.wfn** file, displaying text-based configuration using text editor.

In the subsequent section, we have analyzed a range of examples to elucidate the procedures required to configure QTAIM and NCI analyses utilizing the Gaussian 16 and Multiwfn 3.8 software packages, along with several third-party applications, including Visual Molecular Dynamics (VMD), IrfanView, and GNUPlot, which are contingent upon user discretion for their implementation.

3.4 THEORY OF QUANTUM THEORY OF ATOM IN MOLECULES (QTAIM) AND ITS ANALYSIS

3.4.1 Theory of QTAIM

3.4.1.1 What Is QTAIM?

QTAIM stands out as a sophisticated approach in theoretical chemistry, offering a deeper insight into the molecular structure beyond traditional bonding models. Developed by Richard Bader, this theory is rooted in the quantum mechanical description of electron density, providing a detailed map of how electrons are distributed around atoms in a molecule. Instead of visualizing molecules as simple ball-and-stick models, QTAIM dives into the electron density, i.e., a probability density that paints a vivid picture of where electrons are most likely to be found.

3.4.1.2 Understanding Electron Density

QTAIM mostly revolves around the concept of electron density, $\rho(r)$, derived from a molecule's wavefunction. This electron density isn't uniform or flat; it has peaks around the nuclei and valleys in less electron-dense regions. This landscape of electron density is the foundational map QTAIM uses to explain and predict molecular behavior and interactions.

Critical points in QTAIM are the signposts in the electron density landscape, marked where the gradient of the electron density ($\nabla\rho$) vanishes. These points are characterized by the rank and signature of the Hessian matrix of ρ, which reflects the local curvature:

Nuclear Critical Point (NCP) (3, −3): Here, electron density reaches its peak at the nuclei, showing a local maximum.

Bond Critical Point (BCP) (3, −1): Indicates bonding interactions with a saddle point in electron density between two atoms. This is where the electron density is concentrated along a bond path, facilitating the identification of molecular bonds.

Ring Critical Point (RCP) (3, +1): Found within the planes of atomic rings, highlighting the cyclic nature of the electron paths.

Cage Critical Point (CCP) (3, +3): Represents a local minimum in electron density, typical of enclosed spaces within molecules.

3.4.1.3 Interpreting Molecular Interactions through QTAIM

The real power of QTAIM is in its ability to detail molecular interactions through these critical points:

- **Bond Paths and Orders**: By following the paths of maximum electron density (bond paths) and examining the properties at BCPs, QTAIM assesses bond strength and character.

- **Laplacian of Electron Density**: The Laplacian at a BCP helps decipher the bond's nature, i.e., positive values suggest weaker, ionic or hydrogen bonds, while negative values indicate stronger, covalent bonds.
- **Atoms in Molecules**: QTAIM uniquely defines atoms in a molecule not by arbitrary boundaries but by regions enclosed by zero-flux surfaces in electron density. Each atom thus has its own distinct region, making it easier to analyze its properties independently.

3.4.1.4 Applications of QTAIM

QTAIM isn't just a theoretical curiosity; it has significant practical implications across chemistry and materials science:

- **Predicting Chemical Reactivity**: QTAIM's detailed mapping of electron density and critical points provides insights into potential reactive sites and reaction mechanisms.
- **Material Design**: Understanding how electrons are distributed in materials helps in designing substances with desired properties, crucial in catalyst design and semiconductor manufacturing.
- **Biological Molecule Analysis**: In biochemistry, QTAIM aids in understanding complex molecular interactions within large biological molecules, influencing drug design and protein engineering.

3.4.2 QTAIM Analysis Using Phenol Dimer as Example

The phenol dimer system was chosen for this calculation. The molecules were constructed in **GaussView 6.0** and underwent geometry optimization with **Gaussian 09** at the **B3LYP/6–311 G*** theoretical level. The output files were used to perform single-point energy calculations and generate **.wfn** file, which is a precursor for performing these analyses dedicated to understand the bonding analysis within or between systems. The two systems used for these studies were phenol dimer and water dimer.

Step 1: Perform a single-point energy calculation on optimized phenol dimers and generate **.wfn file** for **AIM analysis** as mentioned in **Section 3.3.1** above

Step 2: Move **AIM batch file** and **text file** to the folder containing **Multiwfn** executable file (Figure 3.6)

Step 3: Use the **.wfn** file generated for the system to be analyzed and transfer it into the directory where Multiwfn software is installed.

Step 4: Open AIM.bat (Batch file) as an administrator and modify the path of input file. In the second line, modify the path to the actual path where VMD package is installed (Figure 3.7).

Step 5: Double click on **AIM.bat** file and run the batch script, which will allow Multiwfn to perform topology analysis of electron density

Step 6: Batch file will be generating **.pdb** files and will be directly generated into the same folder where **VMD** software package is installed. The **.pdb**

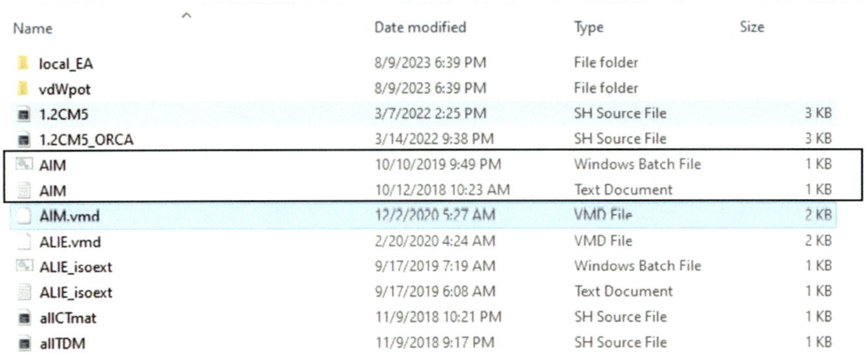

FIGURE 3.6 Screenshot showing AIM batch file and text file setup in the Multiwfn directory for QTAIM analysis.

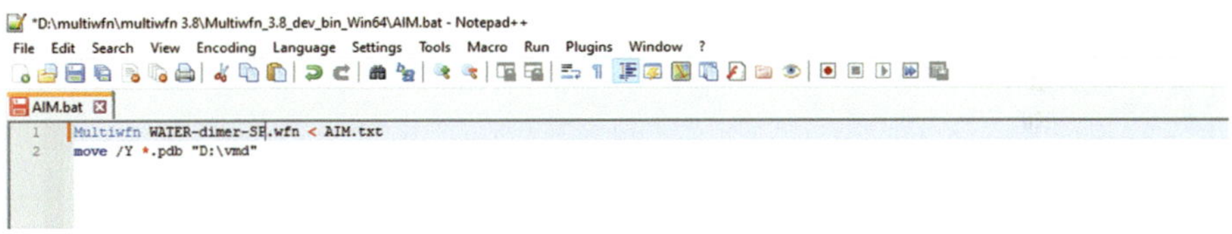

FIGURE 3.7 Example of an **AIM.bat** file setup, used to execute topological analysis of electron density in Multiwfn.

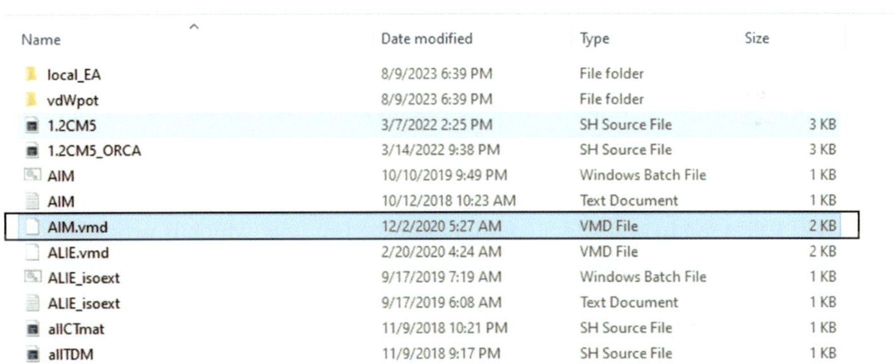

FIGURE 3.8 Screenshot of the **AIM.vmd** script file within the Multiwfn directory, used for visualizing atomic interactions.

files gets automatically generated that contain critical points, molecular structure, and paths.

Step 7: From Multiwfn scripts subdirectory transfer **AIM.vmd** special script to VMD folder (Figure 3.8).

```
###########################################################################
# DESCRIPTION:
#
# VMD startup script. The commands here are executed as soon as VMD starts up
###########################################################################

# turn on lights 0 and 1
light 0 on
light 1 on
light 2 off
light 3 off

# position the stage and axes
axes location lowerleft
stage location off

# position and turn on menus
menu main     move 5   225
menu display  move 395 30
menu graphics move 395 500
menu color    move 125 225
menu files    move 125 325
menu labels   move 125 525
menu render   move 125 525

menu main     on
#menu display  on
#menu graphics on
#menu color    on
#menu labels   on
#menu render   on
#menu files    on

# start the scene a-rockin'
# rock y by 1

# User's default values for newly loaded molecules/reps, etc
#  mol default color {colorID 5}
#  mol default style {NewCartoon 0.300000 6.000000 4.100000 0}
#  mol default selection {name CA}
#  mol default material {Transparent}
proc aim{} {source AIM.vmd}
```

FIGURE 3.9 Example of **vmd.rc** script file modifications, showing how to integrate AIM topology mapping with VMD software.

Step 8: Edit vmd.rc and add a command in the last line, which is ***proc aim {} {source AIM.vmd}*** which enables **VMD** software to plot **AIM topology map** by predefined command. The details about the commands can be understood by referring to the Multiwfn manual (Figure 3.9).

Step 9: Double click on **vmd.exe** and run the application

Step 10: Type command "**aim**" to plot **AIM topology** (Figure 3.10)

Step 11: VMD main tab displays the status of critical points, paths, and molecule structure, which can be fully customized (Figure 3.11)

Step 12: After completion "**CPProp.txt**" named file is generated that contain all properties about **energy and critical points** located in **Multiwfn** directory (Figure 3.12).

Step 13: For more details, the user should refer **Section 2.6** of the **Multiwfn** manual available on their website (**sobereva.com/multiwfn/**).

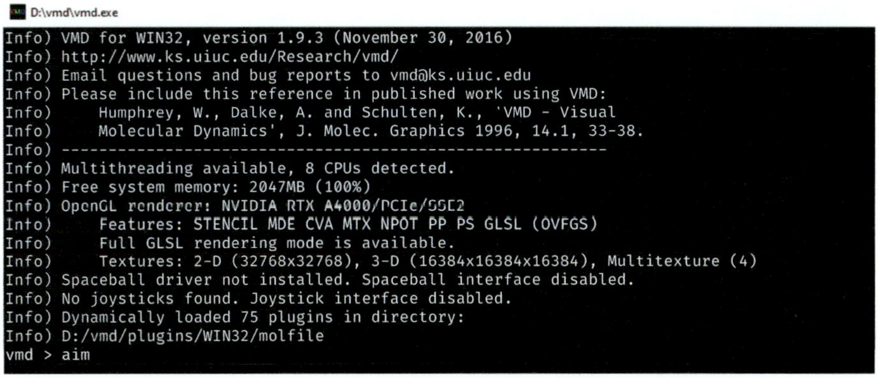

FIGURE 3.10 Screenshot of the VMD command line interface, demonstrating how to execute AIM topology visualization commands.

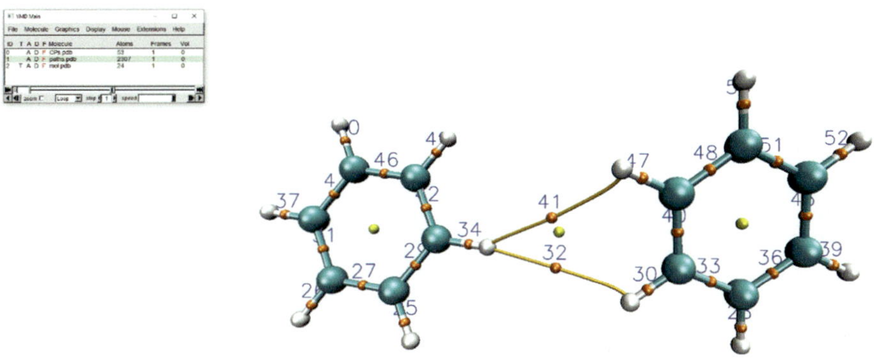

FIGURE 3.11 VMD graphical user interface (GUI) displaying AIM topology of the **phenol dimer**, showing critical points, bond paths, and molecular structures.

3.5 THEORY OF NON-COVALENT INTERACTION (NCI) AND ITS ANALYSIS

3.5.1 Theory of Non-Covalent Interaction (NCI)

3.5.1.1 What Is NCI Analysis?

NCI analysis provides a way to visualize and characterize NCIs in molecular systems. It is particularly useful because it can identify and differentiate between various types of weak interactions that are often challenging to discern using other methods. NCI uses properties derived from the electron density and its derivatives to map out and categorize these interactions.

```
CPprop - Notepad
File Edit Format View Help
--------------     CP   38,     Type (3,-1) ---------------
Connected atoms: 10(C )   --     6(C )
Position (Bohr):      3.870832647119  -0.604116655806 0.427427757254
Position (Angstrom): 2.048356424075  -0.319684766979 0.226185028446
Density of all electrons: 0.3127088739E+00
Density of Alpha electrons: 0.1563544369E+00
Density of Beta electrons: 0.1563544369E+00
Spin density of electrons: 0.0000000000E+00
Lagrangian kinetic energy G(r): 0.1106876629E+00
G(r) in X,Y,Z: 0.4054707827E-01 0.2266639171E-01 0.4747419287E-01
Hamiltonian kinetic energy K(r): 0.3307044715E+00
Potential energy density V(r): -0.4413921344E+00
Energy density E(r) or H(r): -0.3307044715E+00
Laplacian of electron density: -0.8800672348E+00
Electron localization function (ELF): 0.9331707157E+00
Localized orbital locator (LOL): 0.7889010189E+00
Local information entropy: 0.1833957809E-01
Interaction region indicator (IRI): 0.6462030921E-15
Reduced density gradient (RDG): 0.1000000000E+03
Reduced density gradient with promolecular approximation: 0.1000000000E+03
Sign(lambda2)*rho: -0.3127088739E+00
Sign(lambda2)*rho with promolecular approximation: -0.2190452284E+00
Corr. hole for alpha, ref.: 0.00000  0.00000  0.00000 : -0.5491570956E-03
Source function, ref.: 0.00000  0.00000  0.00000 : 0.1777077316E-01
Wavefunction value for orbital    1:  -0.2185127360E-01
Average local ionization energy (ALIE): 0.4873092930E+02
van der Waals potential (probe atom: He): 0.7502179243E+07 kcal/mol
Delta-g (under promolecular approximation): 0.3651464382E+00
Delta-g (under Hirshfeld partition): 0.5577811365E+00
User-defined real space function: 0.7502179243E+07
ESP from nuclear charges: 0.2492879792E+02

Note: Below information are for electron density

Components of gradient in x/y/z are:
 0.4217546451E-16 0.1726049859E-15 -0.2813504638E-16
Norm of gradient is: 0.1798967253E-15

Components of Laplacian in x/y/z are:
-0.3615044961E+00 -0.1822259327E+00 -0.3363368060E+00
Total: -0.8860672348E+00

Hessian matrix:
-0.3615644961E+00 -0.3698917936E+00 0.1997268774E+00
-0.3698917936E+00 -0.1822259327E+00 -0.3455961306E+00
 0.1997268774E+00 -0.3455961306E+00 -0.3363368060E+00
Eigenvalues of Hessian: -0.6754166617E+00 -0.5451460984E+00 0.3404955254E+00
Eigenvectors (columns) of Hessian:
-0.6009482506E+00 -0.6151311759E+00 -8.5103673545E+00
-0.7091274886E+00 0.1157117065E+00 0.6955206725E+00
-0.3687809716E+00 0.7798874518E+00 -0.5057431735E+00
Determinant of Hessian: 0.1253707105E+00
Ellipticity of electron density:  0.238964
```

FIGURE 3.12 Screenshot of **CPProp.txt** output file from Multiwfn, listing details of bond critical points, electron density values, and interaction energies.

3.5.1.2 Key Component: Reduced Density Gradient (RDG)

At the core of NCI analysis is the RDG, a dimensionless scalar field derived from the electron density, $\rho(r)$. RDG is calculated using the formula:

$$\text{RDG}\ (s) = \frac{\nabla\rho(r)}{2(3\pi^2)^{1/3}\rho(r)^{4/3}} \tag{3.1}$$

- $\nabla\rho(r)$ is the gradient of the electron density, indicating how rapidly the electron density changes at a point.
- $\rho(r)^{4/3}$ normalizes this gradient with respect to the local electron density, adjusting for changes in electron density due to the presence of atoms.

3.5.1.3 Significance of RDG in Identifying Interactions

- **Low RDG Values**: Regions where RDG values are low (near zero) typically indicate regions of space where NCIs occur. These are areas where the electron density changes slowly, suggesting stable interactions like hydrogen bonds or stacking interactions.
- **Negative $\nabla^2\rho(r)$ and Low RDG:** Such conditions generally point to attractive interactions, which are stabilizing.
- **Positive $\nabla^2\rho(r)$ and Low RDG:** These indicate nonbonding interactions like steric repulsions.

NCI analysis often uses color-coded scatter plots of RDG versus the sign of the Laplacian of electron density, or versus electron density itself. These plots:

- **Blue Regions**: Indicate strong, attractive interactions (e.g., hydrogen bonds).
- **Red Regions**: Suggest repulsive interactions.
- **Green Regions**: Typically highlight weaker van der Waals forces.

3.5.1.4 Practical Applications of NCI and RDG Analysis

- **Drug Design**: NCI analysis helps in understanding how drugs interact with their biological targets at the molecular level, revealing how different NCIs contribute to drug binding and efficacy.
- **Material Science**: In materials science, understanding these interactions can guide the design of new materials with desired properties, such as increased stability or specific reactivity.
- **Protein Structure and Function**: NCI is crucial in biochemistry for mapping out interaction networks within proteins or between proteins and other biomolecules, influencing their structure and function.

3.5.2 NCI Analysis Using Phenol Dimer as an Example

Step 1: Perform a **single-point energy** calculation on optimized phenol dimers and generate **.wfn** file for AIM analysis as mentioned in **Section 3.3.1** above. Refer **Chapter 4** of **Multiwfn** manual for more information on usage of file type used for this calculation.

Step 2: Move **AIM batch file** and **text file** to the folder containing **Multiwfn** executable file.

Step 3: Use main function "**20**". Various methods are available for visualizing and studying weak interactions (Figure 3.13).

Step 4: Use subfunction "**1**" that belongs to perform **NCI analysis** and properly set up grid for calculating the functions involved in the **NCI analysis**.

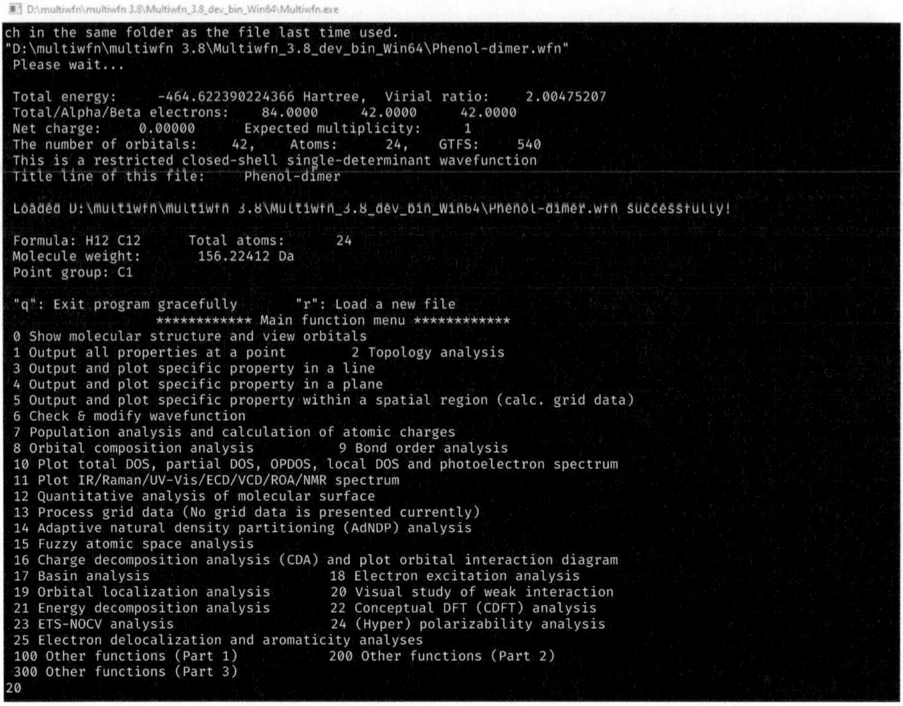

FIGURE 3.13 Screenshot of the Multiwfn terminal, showing the selection process for **Non-Covalent Interaction (NCI) analysis.**

The selection of grid decides the level of accuracy, as the number of generated increases as accuracy increases (Figure 3.14).

Step 5: After the processing is done, you can visualize your scatter map with Multiwfn visualizer, for that Press "**–1**". After visualization is done, you can press right mouse button to close the window (Figures 3.15 and 3.16).

Step 6: Use Subfunction "**7**" that will help you to visualize the **RDG isosurface** map in **Multiwfn** visualizer. Here you have to select the **isosurface** value, which is "**0.5**" to be considered as a standard value (Figure 3.17).

Step 7: Press Enter and you will be able to visualize **isosurface** within the studied system (Figure 3.18).

Step 8: To visualize the **RDG isosurface** within the molecular system with **VMD**, we need to generate .cube files. Press "**3**" and press **Enter** to generate grid data for further processing (Figure 3.19).

Step 9: Two **.cube** files named as **func1 and func2** will be generated in the same **Multiwfn** directory (Figure 3.20).

Step 10: Copy and move these .cube files in **VMD** directory.

Step 11: Copy plotting script named "***RDGfill.vmd***" from "**scripts/Multiwfn**" directory to the **VMD** directory.

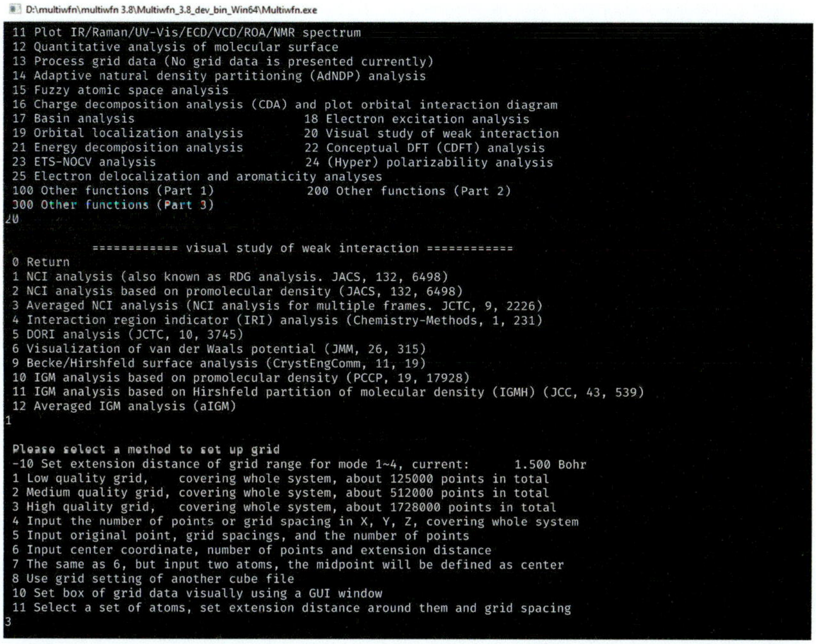

FIGURE 3.14 Screenshot of Multiwfn terminal window configuring grid quality for NCI analysis, highlighting accuracy options.

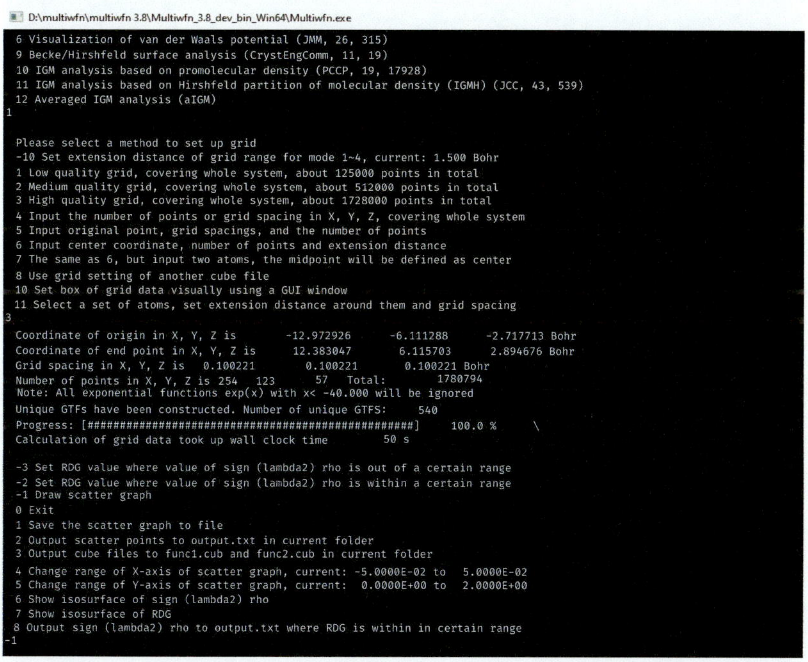

FIGURE 3.15 Multiwfn terminal output displaying scatter graph plotting for non-covalent interactions.

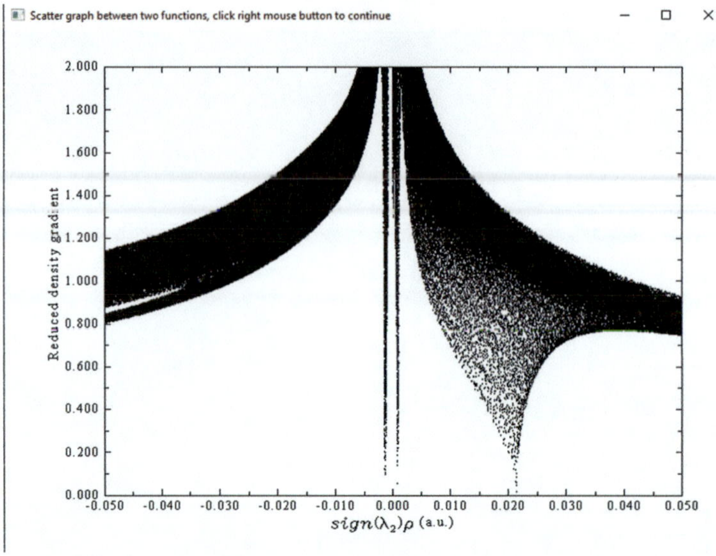

FIGURE 3.16 Example of an **NCI scatter graph** for phenol dimers, generated using Multiwfn.

```
D:\multiwfn\multiwfn 3.8\Multiwfn_3.8_dev_bin_Win64\Multiwfn.exe
 10 Set box of grid data visually using a GUI window
 11 Select a set of atoms, set extension distance around them and grid spacing
3
 Coordinate of origin in X, Y, Z is      -12.972926      -6.111288      -2.717713 Bohr
 Coordinate of end point in X, Y, Z is    12.383047       6.115703       2.894676 Bohr
 Grid spacing in X, Y, Z is      0.100221     0.100221      0.100221 Bohr
 Number of points in X, Y, Z is    254  123   57   Total:        1780794
 Note: All exponential functions exp(x) with x< -40.000 will be ignored
 Unique GTFs have been constructed. Number of unique GTFS:     540
 Progress: [##################################################]     100.0 %
 Calculation of grid data took up wall clock time        50 s

 -3 Set RDG value where value of sign (lambda2) rho is out of a certain range
 -2 Set RDG value where value of sign (lambda2) rho is within a certain range
 -1 Draw scatter graph
 0 Exit
 1 Save the scatter graph to file
 2 Output scatter points to output.txt in current folder
 3 Output cube files to func1.cub and func2.cub in current folder
 4 Change range of X-axis of scatter graph, current: -5.0000E-02 to  5.0000E-02
 5 Change range of Y-axis of scatter graph, current:  0.0000E+00 to  2.0000E+00
 6 Show isosurface of sign (lambda2) rho
 7 Show isosurface of RDG
 8 Output sign (lambda2) rho to output.txt where RDG is within in certain range
-1
 Drawing graph, please wait...

 -3 Set RDG value where value of sign (lambda2) rho is out of a certain range
 -2 Set RDG value where value of sign (lambda2) rho is within a certain range
 -1 Draw scatter graph
 0 Exit
 1 Save the scatter graph to file
 2 Output scatter points to output.txt in current folder
 3 Output cube files to func1.cub and func2.cub in current folder
 4 Change range of X-axis of scatter graph, current: -5.0000E-02 to  5.0000E-02
 5 Change range of Y-axis of scatter graph, current:  0.0000E+00 to  2.0000E+00
 6 Show isosurface of sign (lambda2) rho
 7 Show isosurface of RDG
 8 Output sign (lambda2) rho to output.txt where RDG is within in certain range
7
 Input the value of isosurface, e.g. 0.5
0.5
```

FIGURE 3.17 Multiwfn terminal window showing input parameters for **Reduced Density Gradient (RDG) isosurface** plotting in NCI analysis.

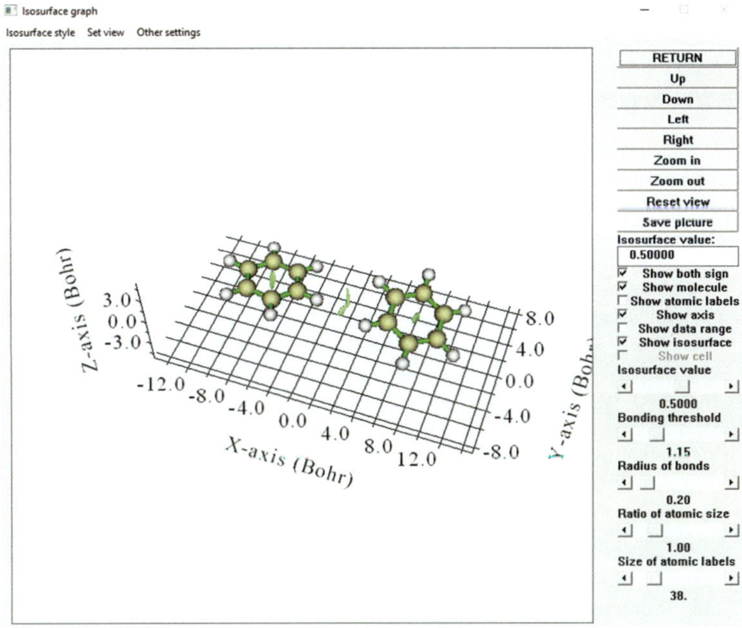

FIGURE 3.18 Multiwfn visualization displaying the **RDG isosurface** for phenol dimers, highlighting non-covalent interactions.

```
D:\multiwfn\multiwfn 3.8\Multiwfn_3.8_dev_bin_Win64\Multiwfn.exe
 1 Save the scatter graph to file
 2 Output scatter points to output.txt in current folder
 3 Output cube files to funci.cub and func2.cub in current folder
 4 Change range of X-axis of scatter graph, current: -5.0000E-02 to   5.0000E-02
 5 Change range of Y-axis of scatter graph, current:  0.0000E+00 to   2.0000E+00
 6 Show isosurface of sign (lambda2) rho
 7 Show isosurface of RDG
 8 Output sign (lambda2) rho to output.txt where RDG is within in certain range
7
 Input the value of isosurface, e.g. 0.5
0.5

 -3 Set RDG value where value of sign (lambda2) rho is out of a certain range
 -2 Set RDG value where value of sign (lambda2) rho is within a certain range
 -1 Draw scatter graph
 0 Exit
 1 Save the scatter graph to file
 2 Output scatter points to output.txt in current folder
 3 Output cube files to func1.cub and func2.cub in current folder
 4 Change range of X-axis of scatter graph, current: -5.0000E-02 to   5.0000E-02
 5 Change range of Y-axis of scatter graph, current:  0.0000E+00 to   2.0000E+00
 6 Show isosurface of sign (lambda2) rho
 7 Show isosurface of RDG
 8 Output sign (lambda2) rho to output.txt where RDG is within in certain range
3
 Exporting ...
 The cube file of sign (lambda2) rho has been exported to func1.cub in current folder
 The cube file of RDG has been exported to func2.cub in current folder
```

FIGURE 3.19 Multiwfn terminal screenshot showing the process for generating **.cube** files, used for RDG and electron density visualization.

FIGURE 3.20 Screenshot showing a set of **.cube files** generated by Multiwfn for phenol dimers.

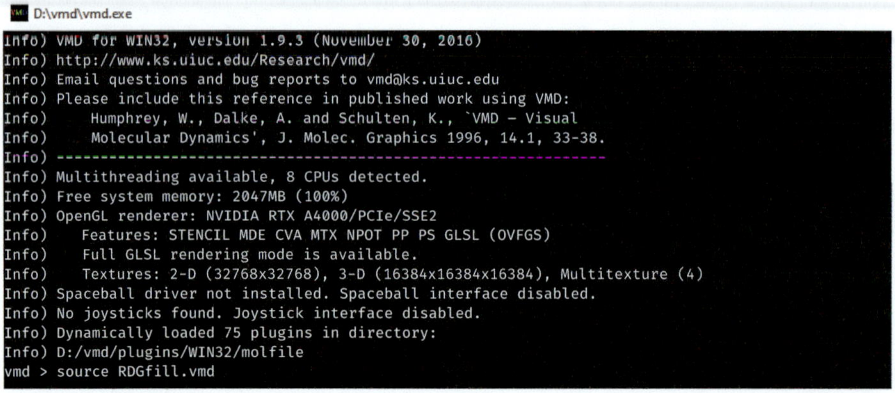

FIGURE 3.21 Screenshot of the VMD terminal, showing command execution for plotting RDG isosurfaces from **.cube** files.

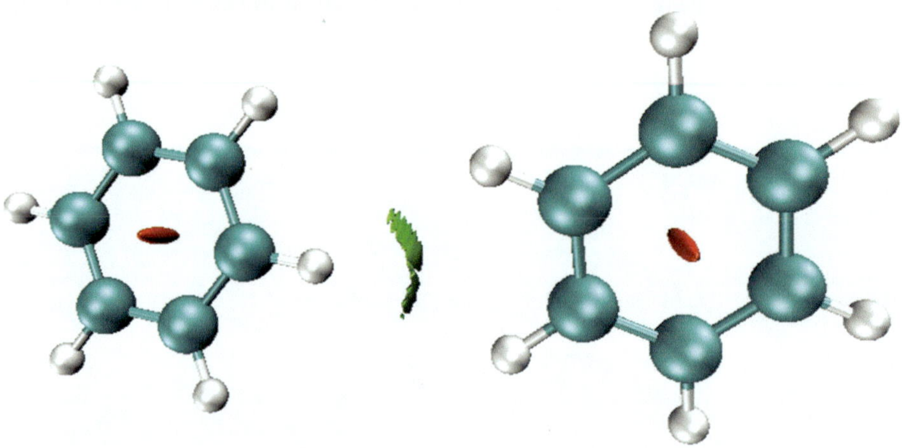

FIGURE 3.22 VMD-rendered **isosurface plot** of non-covalent interactions in phenol dimers, showing weak molecular interactions.

Step 12: Go to **VMD** directory and run **VMD.exe**.

Step 13: Type command in the **VMD** command line "***source RDGfill.vmd***", which will guide the plotting script (Figure 3.21).

Step 13: Press Enter and you will get the isosurface for the phenol dimer system (Figure 3.22).

Step 14: Notice that the **isovalue** effects representation of the colored **RDG** isosurface map, you can slightly adjust it to try to improve the graph. Isovalue of **0.5–0.6** is mostly preferred, for this refer 3.23.1 in manual (Figure 3.23).

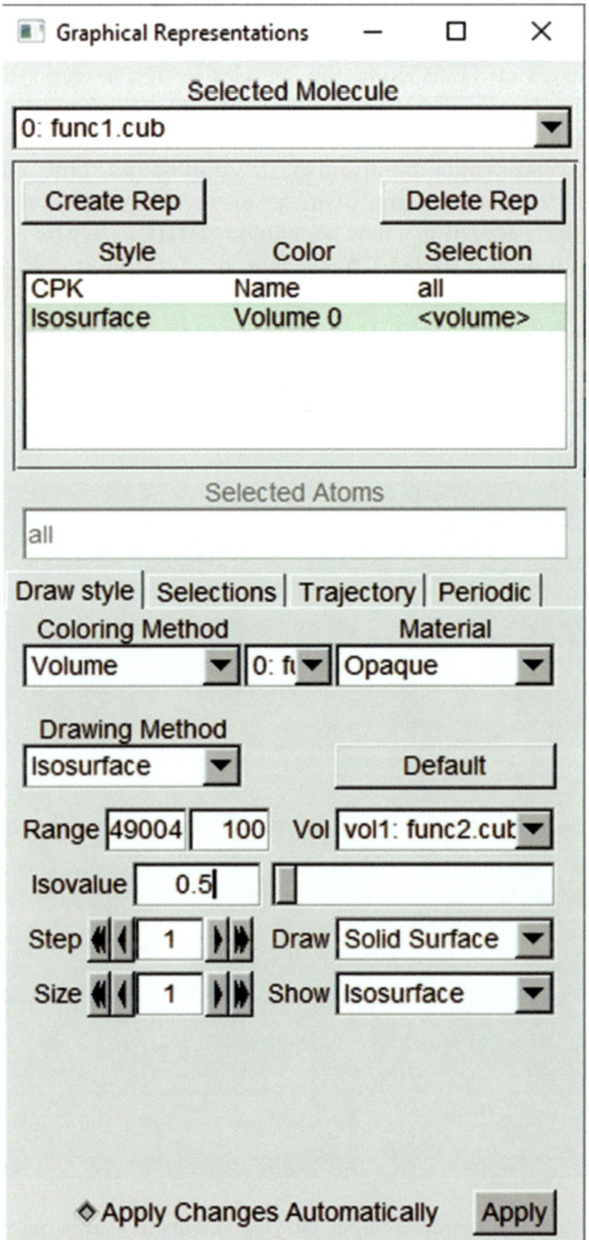

FIGURE 3.23 Screenshot of the VMD interface for modifying **isosurface values**, allowing users to adjust visualization parameters.

Step 15: In order to plot a colored scattered graph by third-party software's, we have to export the scatter points, as a text file. Press "**2**" in the **multiwfn** window and press enter. This will output scatter points to file named "**output.txt**" in the current folder (Figure 3.24).

Step 16: Move the generated "**.txt**" file to "**bin/gnuplot**" directory where **GnuPlot** software package is installed.

Step 17: Next step is to copy the gnuplot script named "**RDGscatter.gnu**" from "**scripts/Multiwfn**" directory to "**bin/gnuplot**" directory (Figure 3.25).

Step 18: Open command prompt and locate **gnuplot** and "**bin**" folder and type command "**RDGscatter.gnu**", which will invoke **gnuplot** to plot the data via the script, generating a new file named "**RDGscatter.ps**".

Step 19: If you have installed **Ghostscript** and **IrfanView**, the generated **.ps** file can be directly visualized. The generated **.ps** file can also be opened by many other programs such as Acrobat; however, the loading can be slow (Figure 3.26).

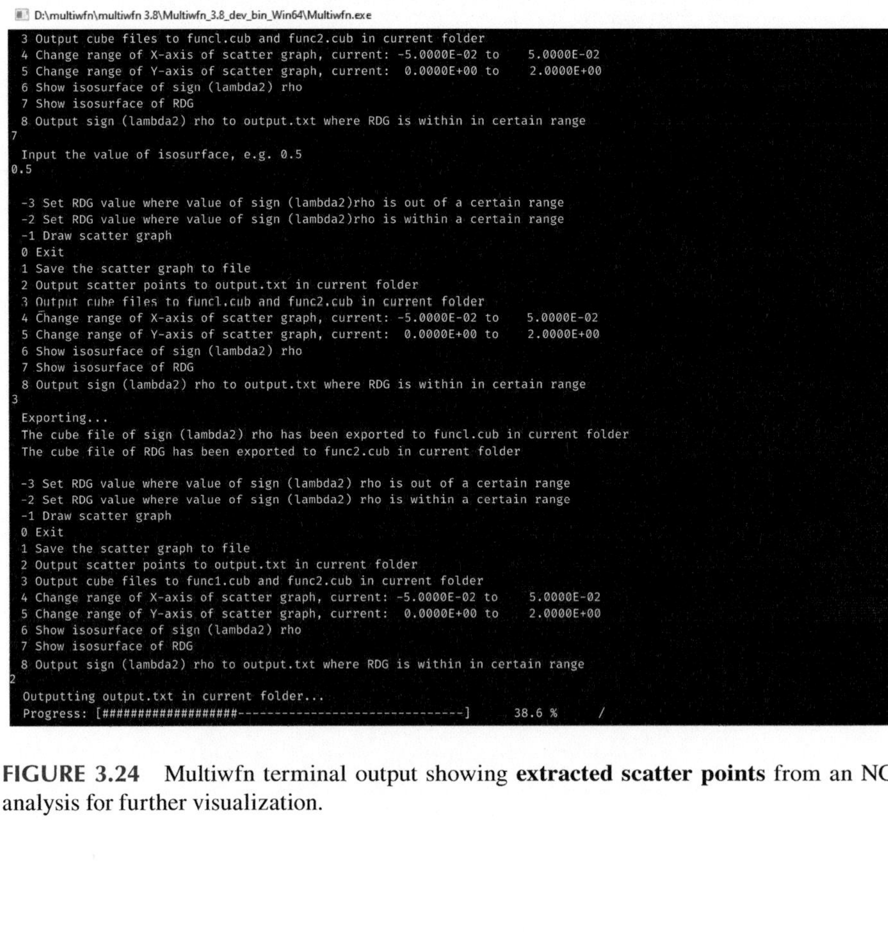

FIGURE 3.24 Multiwfn terminal output showing **extracted scatter points** from an NCI analysis for further visualization.

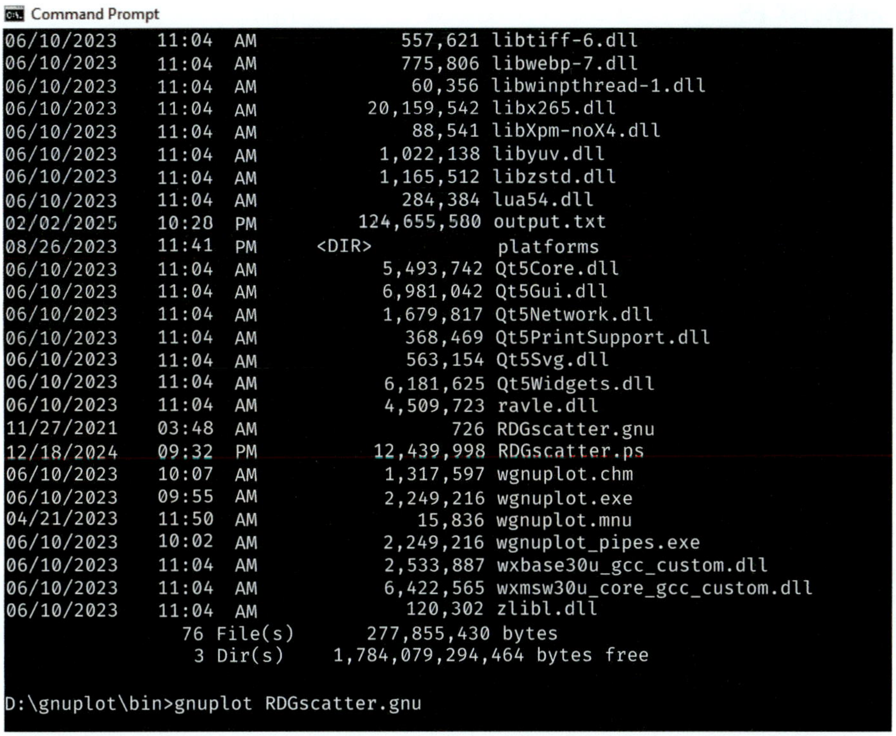

FIGURE 3.25 Screenshot of **Gnuplot** command prompt used for plotting RDG scatter graphs from NCI analysis.

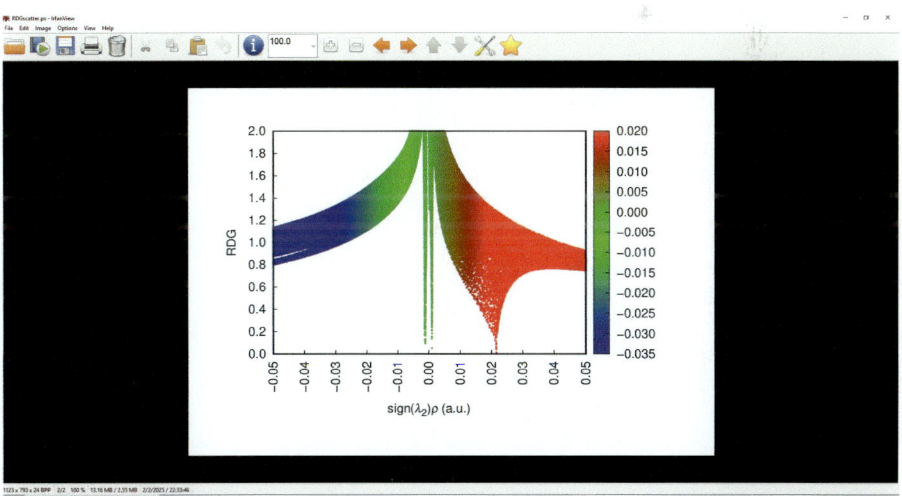

FIGURE 3.26 Final **NCI scatter plot** for phenol dimers, displaying attractive and repulsive non-covalent interactions.

REFERENCES

1. Gilli, G. and P. Gilli, *The nature of the hydrogen bond: outline of a comprehensive hydrogen bond theory*. Vol. 23. 2009: Oxford university press.
2. Latimer, W.M. and W.H. Rodebush, Polarity and ionization from the standpoint of the Lewis theory of valence. *Journal of the American Chemical Society*, 1920. **42**(7) : p. 1419–1433.
3. Arunan, E., et al., Definition of the hydrogen bond (IUPAC Recommendations 2011). *Pure and Applied Chemistry*, 2011. **83**(8): p. 1637–1641.
4. Kollman, P.A. and L.C. Allen, Theory of the hydrogen bond. *Chemical Reviews*, 1972. **72**(3): p. 283–303.
5. Coulson, C., THE HYDROGEN BOND: A review of the present interpretations and an introduction to the theoretical papers presented at the Congress. *Hydrogen Bonding*, 1959: p. 339–360.
6. Grabowski, S.J., *Hydrogen bonding: new insights*. Vol. 3. 2006: Springer.
7. Gilli, P., et al., A novel approach to hydrogen bonding theory. *Advances in Molecular Structure Research*, 1996. **2**: p. 67–102.
8. Lu, T. and F. Chen, Multiwfn: A multifunctional wavefunction analyzer. *Journal of Computational Chemistry*, 2012. **33**(5): p. 580–592.

4 QM/MM Methods for Solvent Studies

4.1 INTRODUCTION

Solvation is a fundamental phenomenon that significantly influences the chemical behavior of molecules in solutions. It affects reaction mechanisms, thermodynamics, and kinetics by altering the stability and reactivity of solutes. Computational solvation studies have become indispensable for understanding these effects at a molecular level. This chapter focuses on three primary computational approaches to modeling solvation: explicit, implicit, and hybrid methods such as the ONIOM model.

Explicit solvation methods simulate each molecule of the solvent separately, giving a full picture of how the solvent interacts with the solute. These methods are very good at modeling the area around the solute because they take into account specific interactions like hydrogen bonding and van der Waals forces. Techniques such as molecular dynamics and Monte Carlo simulations fall under this category. However, the computational cost is significant due to the large number of particles and interactions that must be considered.

Implicit solvation methods simplify the solvent environment by treating it as a continuous dielectric medium. The Polarizable Continuum Model (PCM) and the Generalized Born (GB) approximations do away with discrete solvent molecules and instead use a solvent field that impacts the solute. This approach reduces computational demands while capturing the bulk properties of the solvent. Even though implicit methods aren't as detailed, they are useful for looking at bigger systems or when studying specific interactions between solvents and solutes isn't the main goal.

The ONIOM model (Our own N-layered Integrated Molecular Orbital and molecular Mechanics) is an example of a hybrid method that takes the best parts of both explicit and implicit approaches. ONIOM lets you treat the system in more than one way. A high-level quantum mechanical method is used to model the area of interest, which includes the solute and immediate solvent shell. A lower-level method or a continuum is used to model the environment around it. This enables accurate modeling of critical interactions without prohibitive computational costs.

In the sections that follow, we'll talk more about implicit, explicit, and ONIOM methods, including how they can be used in real-world solvation research today. By contrasting explicit, implicit, and hybrid models, we hope to give researchers a full picture of their pros and cons, which will help them choose the best method for their specific solvation studies.

DOI: 10.1201/9781032694733-4

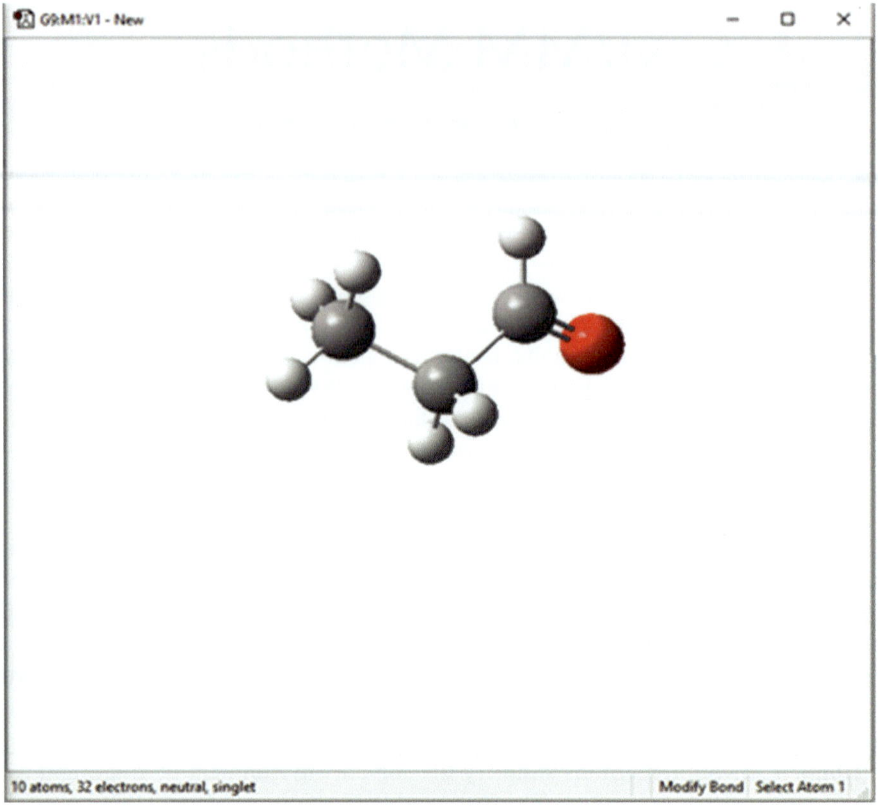

FIGURE 4.1 Illustration of optimized propanal molecules in Gauss View.

4.1.1 Implicit Model of Solvation

To understand the *implicit model* of solvation in GaussView 6.0.16 [1]. Here, we will illustrate the steps to apply the solvation model, such as SMD, on a small molecule such as propanal (Figure 4.1).

Step 1: First, draw the most appropriate propanal structure and choose the job type (Figure 4.2)**.**
Step 2: Select the **Method tab** in the Gaussian [2] calculation setup and select the basis set theory (Figure 4.3).
Step 3: The solvation panel allows you to specify if the calculation is to be performed in the solvent phase rather than in the gas phase. You can select the model from the top-down list. It is illustrated in Figure 4.4.

The model field will allow you to specify the solvation model (the default is IEFPCM). Some solvation models may present different/additional fields for their required parameters.

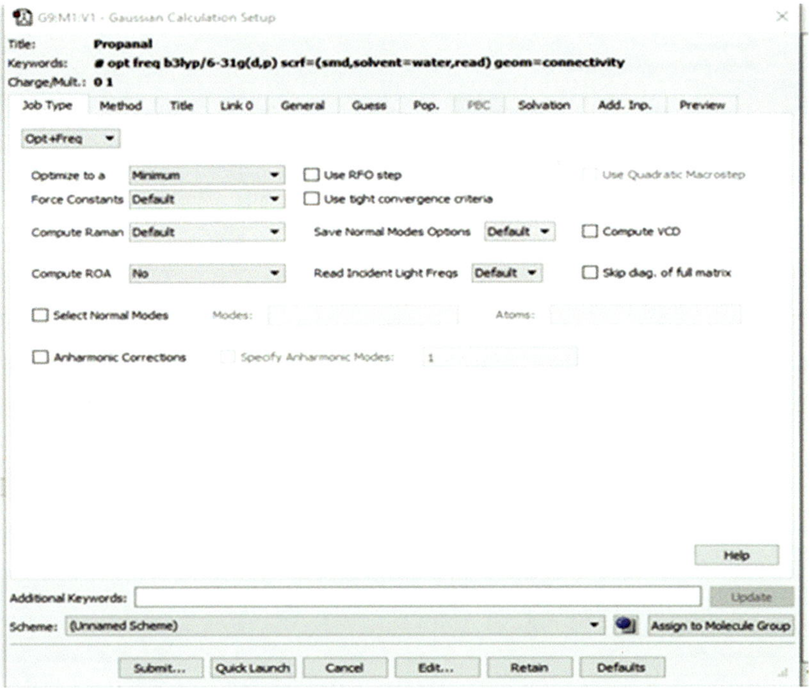

FIGURE 4.2 Illustration of optimized propanal molecule calculation setup with details of job-type dialogue.

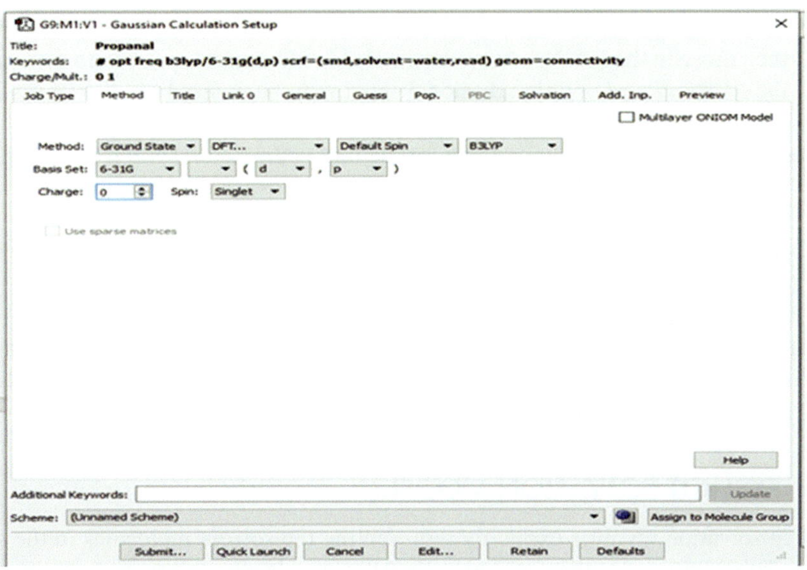

FIGURE 4.3 Optimized propanal molecule is illustrated, with specifics on the Method tab.

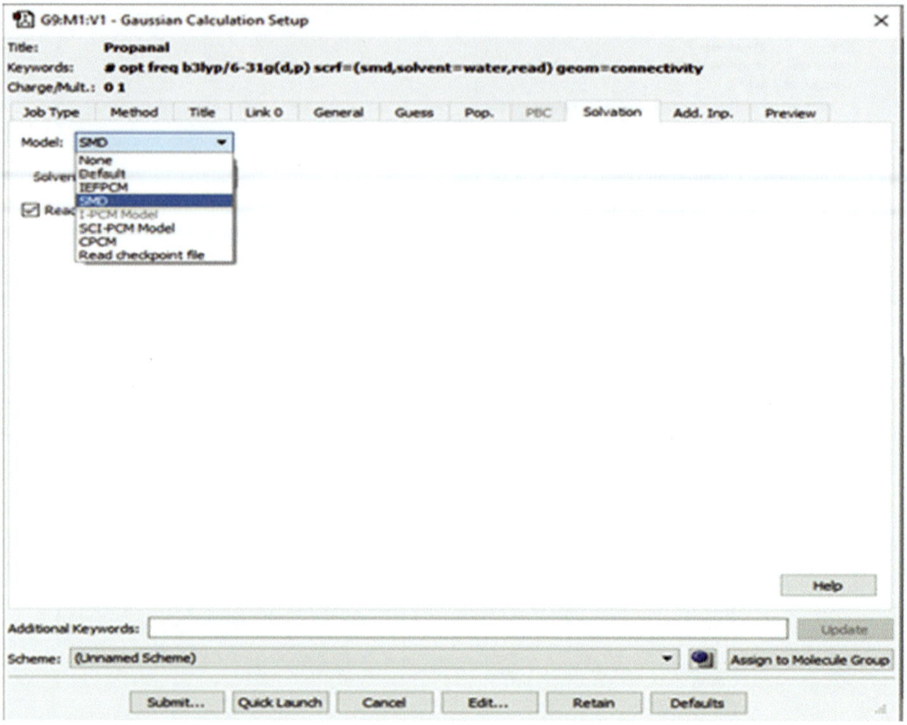

FIGURE 4.4 Illustration demonstrating the steps of choosing a solvation model.

Step 4: You can specify the specific solvent by selecting it from the corresponding top-down menu. Use the other selection to select a solvent other than those in the list. You may use the Read Additional Input box to generate the *SCRF=Read* option; additional input may be entered via the Gaussian calculation setup (Figure 4.5).

Step 5: You can check the preview for the input parameters, and if anything is missing, you can make the changes in the respective tabs. The job card will look like this.

#opt freq b3lyp/6-31g(d,p) scrf=(smd, solvent=water, read) geom=connectivity

If everything looks fine, then save the file and submit the job (Figure 4.6).

Calculations from the log file generated after completion of job, yield gas and solvation phase energies. You can calculate the Gibbs energy of solvation by subtracting the gas phase energy from the solvation phase energy.

4.2 EXPLICIT SOLVATION MODEL

Explicit solvation involves explicitly representing the solvent molecules in the simulation. This implies that the simulation models the solvent as a group of distinct molecules that engage in direct interactions with both the solute and each other. We treat

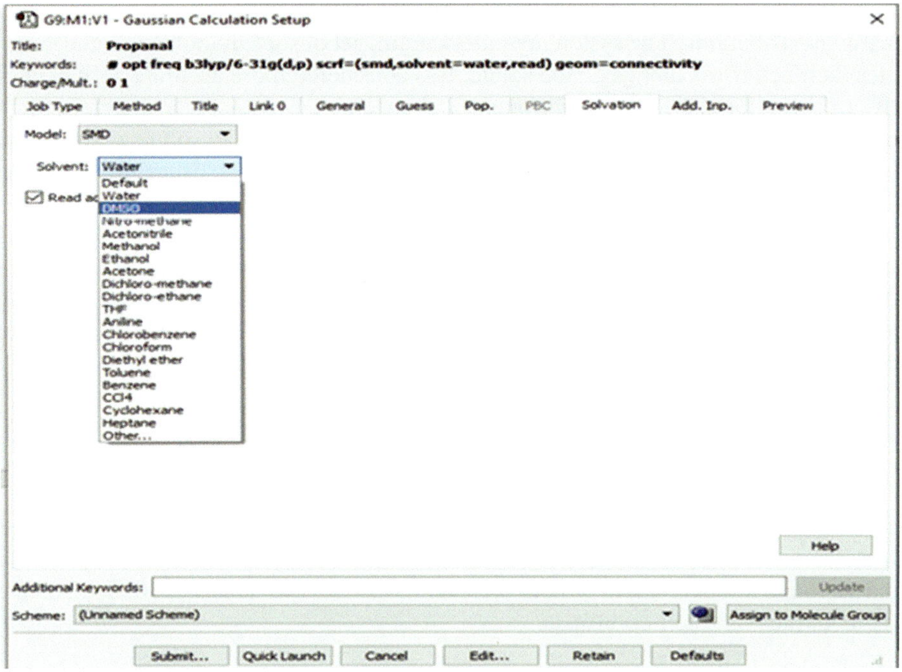

FIGURE 4.5 Illustration of selecting the solvent for different solvation models.

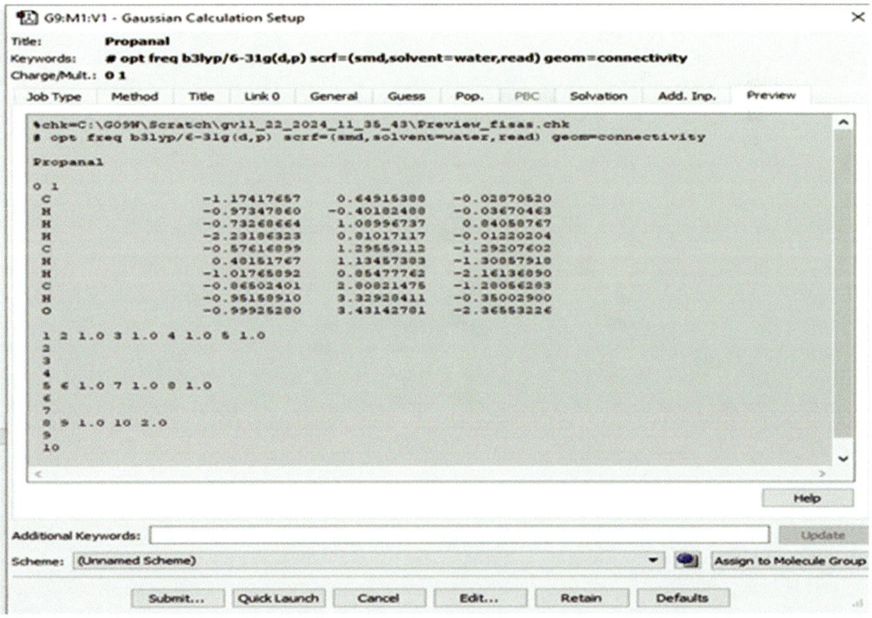

FIGURE 4.6 Illustration of the preview dialogue box before submitting the job.

each solvent molecule as an individual entity with defined positions, interactions, and energy contributions. The system includes the full set of solvent molecules and simulates their behaviors alongside the solute. It is considered more accurate, as it explicitly accounts for the local environment and specific solvent–solute interactions. It can model complex solvation effects, such as hydrogen bonding, solute aggregation, and solvent structure. However, it is computationally expensive, especially for large systems, because the number of solvent molecules can be very large. It also requires more resources and time to perform simulations, particularly for large-scale systems with many solvent molecules. Explicit solvation requires adding solvent molecules to the input structure manually or by random packing. For larger systems or more complicated solvent effects, implicit solvation (like PCM or IEFPCM) may be more useful, as it doesn't requires explicitly adding solvent molecules.

4.2.1 Using Explicit Solvation for Ethanol Solvation Using the Water Cluster

Let's say you want to study a **small solute** (like a molecule of ethanol) surrounded by a cluster of **water molecules** (explicit solvation). You'd place ethanol in the center and surround it with a small number of water molecules (perhaps 5 or 10), then create the input file with their coordinates.

Step 1: Draw ethanol molecules surrounded by the 7 water molecules to study the explicit solvation by the water molecules (Figure 4.7).
Step 2: Fill in the method card details as given below. Complete all the details in other tabs and provide the title of the job (Figure 4.8).
Step 3: The preview will look like the ones given below (Figure 4.9).

FIGURE 4.7 Illustration of propanal molecules surrounded by water molecules for explicit calculation.

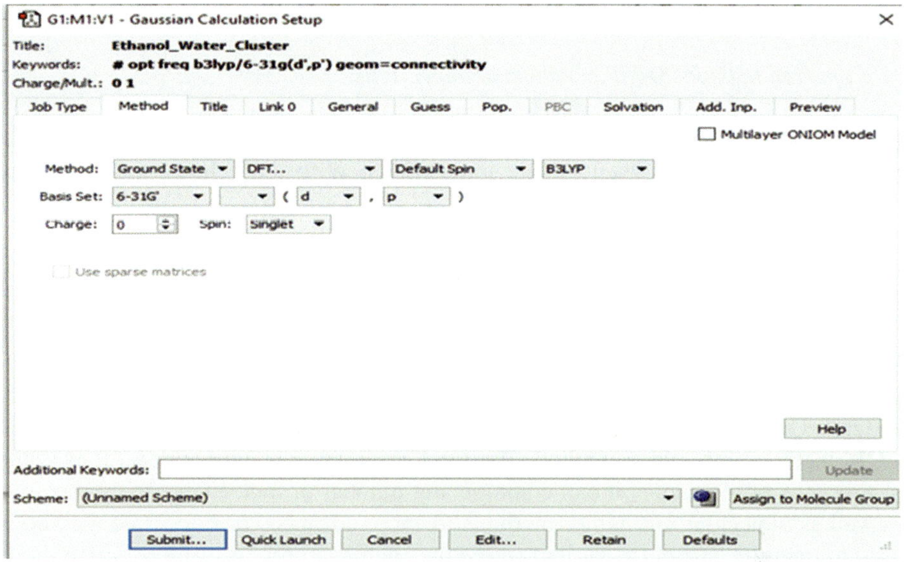

FIGURE 4.8 Illustration of method card details for propanal-water cluster.

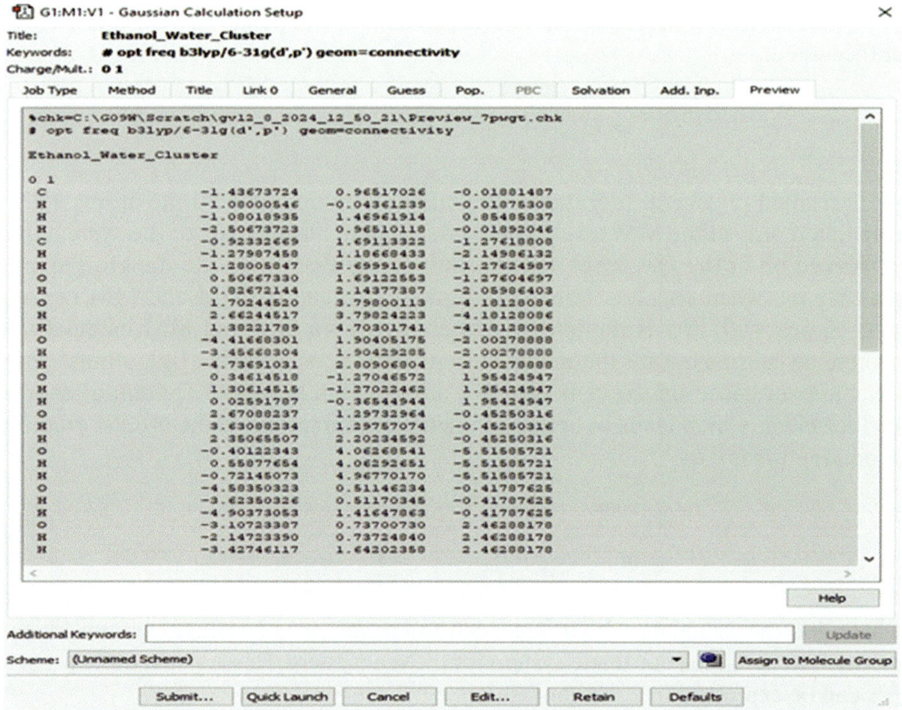

FIGURE 4.9 Preview of propanal-water cluster submission job.

Job route keywords:

 # opt freq b3lyp/6-31g(d',p') geom=connectivity

After running the job, this type of calculation explicitly accounts for the local environment and specific solvent–solute interactions. With the help of water–ethanol cluster systems, it can model complicated solvation effects such as hydrogen bonding, solute aggregation, and solvent structure.

4.3 ONIOM BACKGROUND

Hybrid techniques that combine two or more computational methods in one calculation allow the accurate exploration of the chemistry of very large and complex systems. Quantum mechanical (QM)/Molecular Mechanics (MM) methods combine a QM method with a MM method. The ONIOM scheme is a more general example of this type of technique; it can combine any number of molecular orbital methods as well as molecular mechanics methods. A more advanced and accurate method is used to treat the part of the system where the chemical process or more active site is located, such as where bond breaking and formation happen. The lower-level method is used to treat the rest of the system. The QM/MM method has been useful for studying enzyme reactions. The active site is studied using a high-level method, most often DFT, and the protein environment is studied using molecular mechanics. In a two-layer ONIOM calculation, three independent calculations yield the total energy of the system.

$$E^{\text{ONIOM(QM:MM)}} = E^{\text{QM}}_{\text{Model}} + E^{\text{MM}}_{\text{Real}} - E^{\text{MM}}_{\text{Model}} = E^{\text{High}}_{\text{Model}} + E^{\text{Low}}_{\text{Real}} - E^{\text{Low}}_{\text{Model}} \tag{4.1}$$

As described in several literatures, the real system contains all the atoms and is calculated only at the MM level. The model system has the part of the system that is worked on at the QM level and the link atoms that cover the dangling bonds that happen when covalent bonds are broken between the QM and MM regions (see Figure 4.10). To evaluate ONIOM energy, both QM and MM calculations need to be carried out for the model system. This is because the link atoms' positions are set based on the real system's atoms. This makes the potential energy surface clear, which leads to better geometry optimization. The ONIOM gradient is obtained from

$$\frac{\partial E^{\text{ONIOM}}}{\partial q} = \frac{\partial E^{\text{high}}_{\text{model}}}{\partial q} \cdot j + \frac{\partial E^{\text{low}}_{\text{real}}}{\partial q} - \frac{\partial E^{\text{low}}_{\text{model}}}{\partial q} \cdot j \tag{4.2}$$

where J is the Jacobian, which is needed to convert the coordinate system for the model system to the coordinate system for the real system. Hessian and other properties can be expressed in a similar fashion.

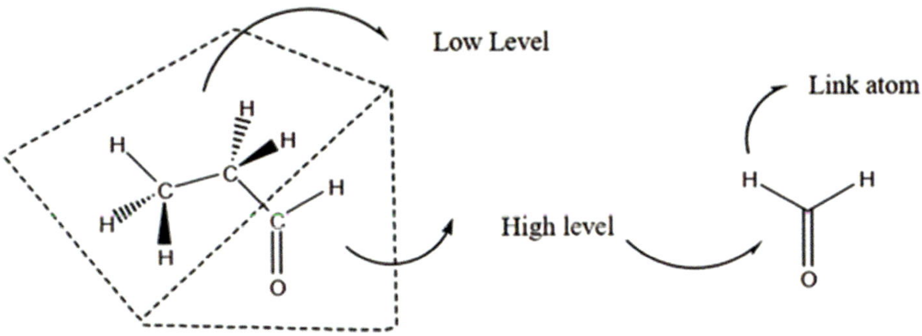

FIGURE 4.10 Illustration of two-layer model of propanal for ONIOM calculation.

4.3.1 ONIOM Method

Let's start by figuring out the frequency and energy of the propanal carbonyl functional group in order to show how the ONIOM method can be used. When doing an ONIOM calculation, link atoms are used to model atoms at the model/real system boundary. These are heavy atoms in the real systems that are connected to atoms in the small model (SM) system. The calculations are run on the SM system. For instance, the computations typically replace a bonded carbon atom within the SM system with hydrogen. The following diagram illustrates link atoms, using ethane as an example (Figure 4.10):

In the (double-layered) ONIOM approach, we divide the molecule into two pieces. The region of most interest (containing the carbonyl group here) is at a high level, and the rest of the molecules are low level. If we are interested in the CO bond length and frequency, we can approximate the energy as

$$E(\text{ONIOM}) = E\left(\text{propanal / low level}\right) + E\left(\text{formaldehyde / high level}\right) \\ - E\left(\text{formaldehyde / low level}\right)$$

We can use this compound formula for the energy to optimize the structure and to calculate the force constants. The positions of all the atoms are optimized for the complete structure, while the position of the link atom is somewhat arbitrary and uses a protocol developed in the group of Keiji Morokuma, the inventor of ONIOM, at Emory University [3].

In an ONIOM calculation, the molecular system being analyzed is delineated into two or three regions, commonly known as "layers". These layers are represented using progressively precise model chemistries:

1. The uppermost layer is the smallest and is processed using the most precise advanced technique. In this region, bond formation and breaking occur. The layer is referred to as the SM system. In a two-layer ONIOM calculation, this is commonly referred to as the model system.
2. The low layer encompasses the complete molecule within a two-layer ONIOM framework. The computation for this area pertains to the environmental impacts of the molecular surroundings on the specified site (i.e., the

upper layer). The lower layer is generally addressed using a cost-effective model chemistry, which may include molecular mechanics, a semi-empirical approach, or a budget-friendly ab initio method like Hartree–Fock utilizing a modest basis set.

3. In a three-layer ONIOM model, a central layer is established, and it is addressed using a method that possesses an accuracy level that is intermediate between the high-level method and the low-level method. This layer encompasses a more extensive subset of the complete molecule compared to the upper layer. The model represents the electronic effects of the molecular environment on the upper layer. The integration of this layer with the lower layer is designated as the intermediate model (IM) system. The complete structure is designated as the real system.

4.3.2 Setting-Up ONIOM Calculation for the Two-Layer Propanal

Step 1: First draw propanal molecules using the *Builder* tab of GaussView 6.0.16, as shown in Figure 4.11.

Step 2: Then select the **Multilayer ONIOM Model** in the calculation setup under the method card. On the **Method** tab, check **Multilayer ONIOM Model** as shown by pointing the arrow (Figure 4.12).

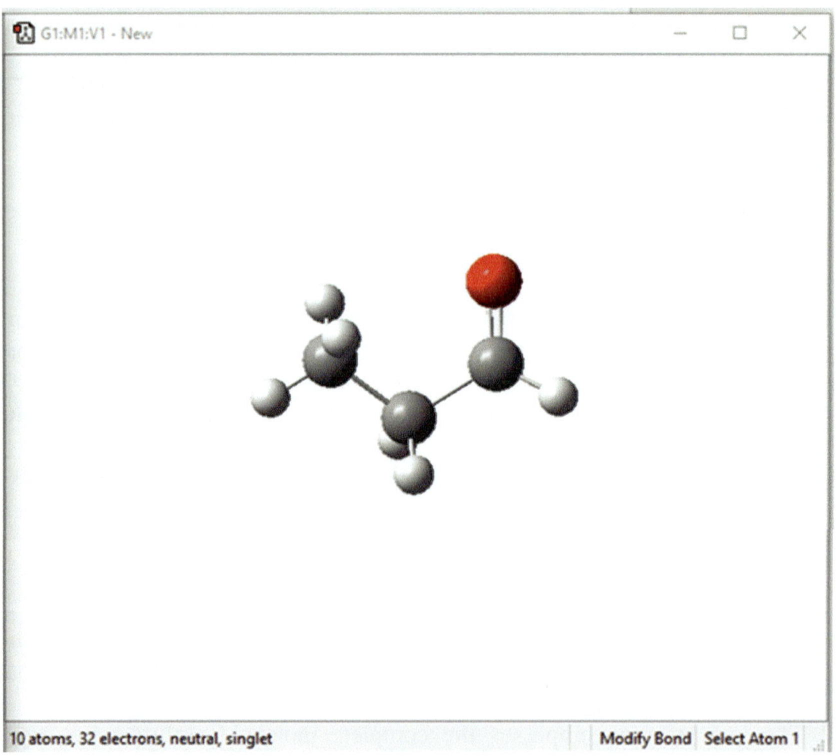

FIGURE 4.11 Propanal molecule as drawn in Gauss View for ONIOM calculations.

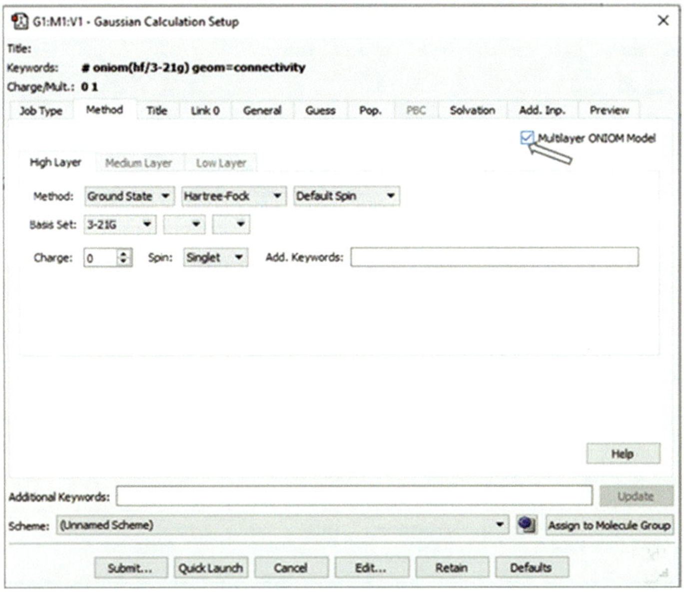

FIGURE 4.12 Checkbox for multilayer for the ONIOM model.

Step 3: We now must assign the atoms to the low and high regions or layers. This is done using the layer selection tool under the edit menu of GaussView. Choosing the Edit=>Select Layer menu item or clicking on the Select Layer button opens the Layer Selection Tool (pointed by the arrow). This window is used to assign atoms to ONIOM layers. We illustrate it in Figure 4.13.

Step 4: Three tabs appear (**High**, **Medium**, and **Low**) in the layer selection tool, which you can use to select the methods for treating each layer of atoms (Figure 4.14)

We are about to place the selected atoms into the low layer. This dialog allows you to assign atoms to layers for ONIOM calculations. The different display formats indicate different layers. Here, you can select several atoms for layer assignment, or you can select all for layer selection. For this case, I have pressed Select All as shown (pointed by the arrow) in Figure 4.15.

After pressing Select All (pointed by the arrow), it will look as shown in Figure 4.16.

Once you press the apply button, the molecule transitions to a wireframe display, indicating that you have selected a low layer in the display tab. Each of the atoms is displayed by an atom number activated by pressing the number in view (Figures 4.17 and 4.18).

Step 5: Next go on to the **(a) High Layer** using the layer selection tool, first initialize by **(b) Select None** and then select atoms by clicking on them and holding down the shift key. When you are done selecting the atoms, **(c)** press **Apply** to make the changes. At this point, we have selected the layers (Figures 4.19 and 4.20).

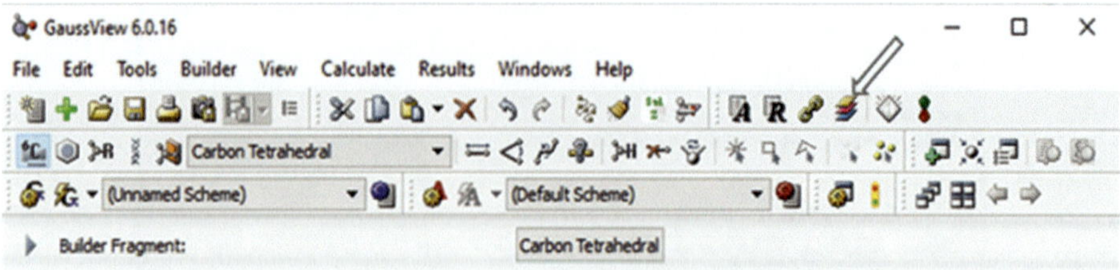

FIGURE 4.13 Layer selection menu via arrow for the multilayer ONIOM model.

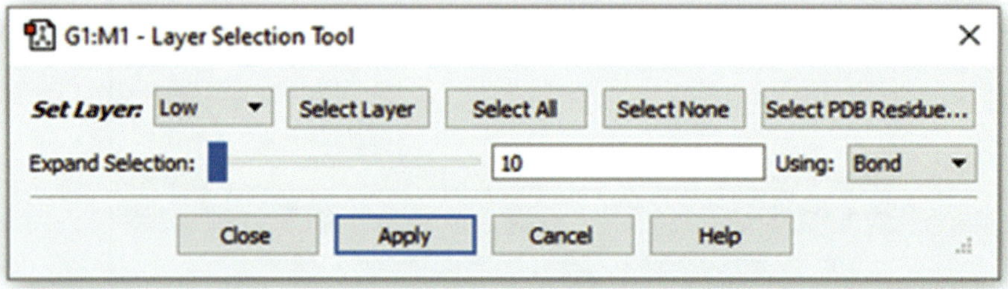

FIGURE 4.14 Layer selection tool with option to choose different layers by using the drop down menu.

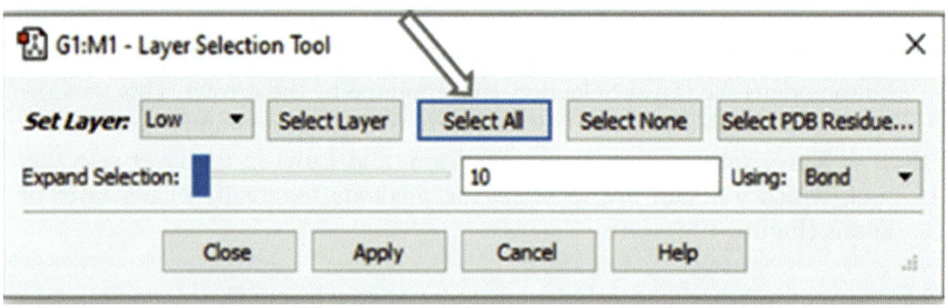

FIGURE 4.15 Select all button of layer selection tool.

Step 6: In GaussView 6.0, the atoms in different layers are rendered differently, so you can check your selection. Under *view/display/format/molecule,* you can change the visualization. I prefer tubes in my low layer in case of a double-layer calculation.

The various supported display types are

Ball & Stick: The molecule appears as a ball-and-stick model, with all types of bonds represented by single sticks.

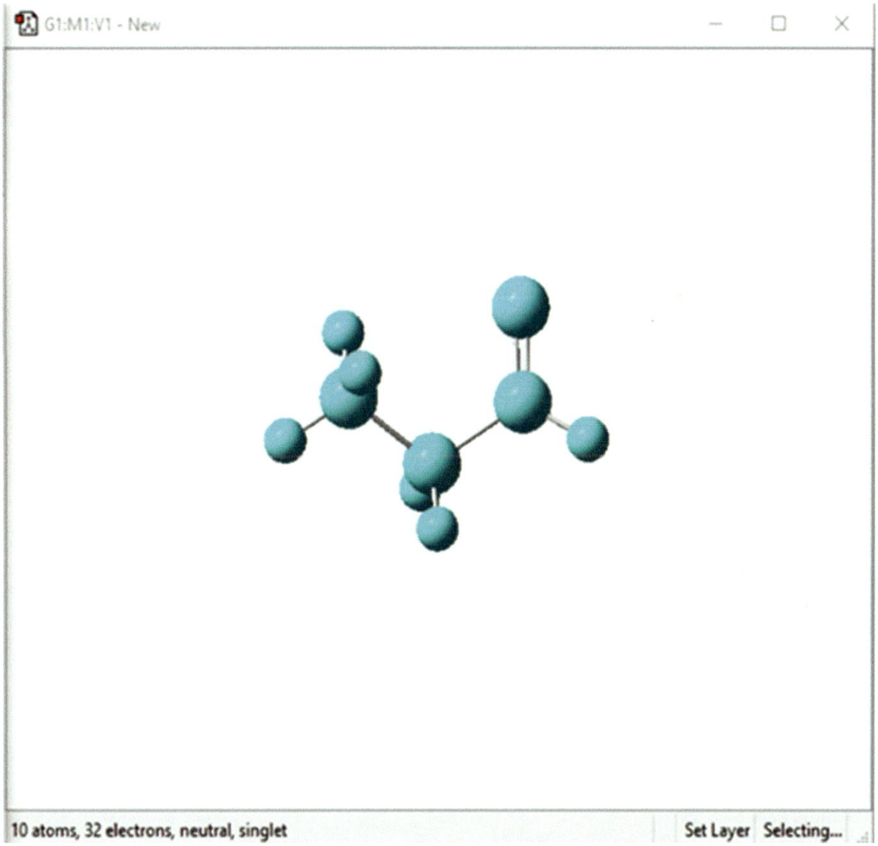

FIGURE 4.16 Selected atoms for the low layer in the ONIOM model.

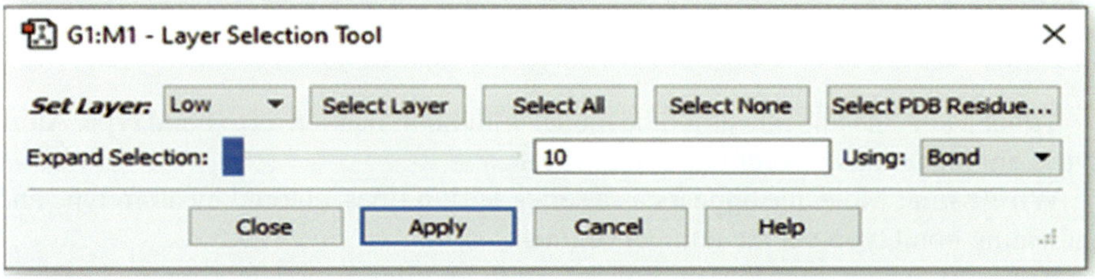

FIGURE 4.17 Apply button of the layer selection tool.

Ball & Bond Type: Molecule appears as a ball and stick model with single bonds represented by single sticks and multiple bonds represented by multiple sticks or, in the case of aromatic systems, by dotted lines. This is the default option.

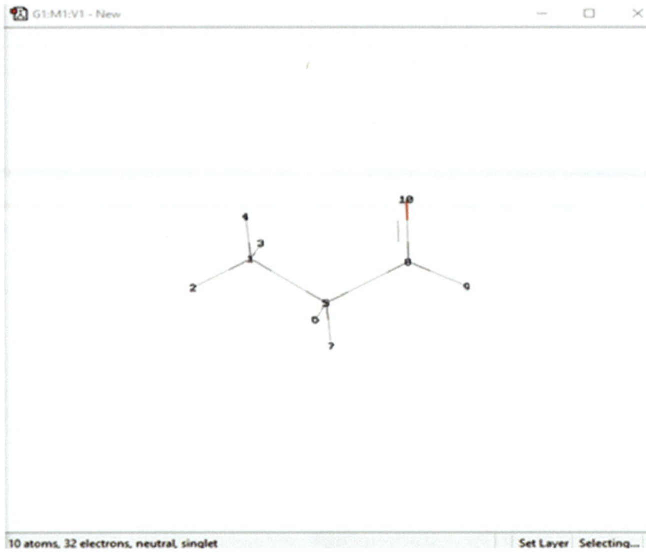

FIGURE 4.18 Wireframe structure of propanal after pressing apply.

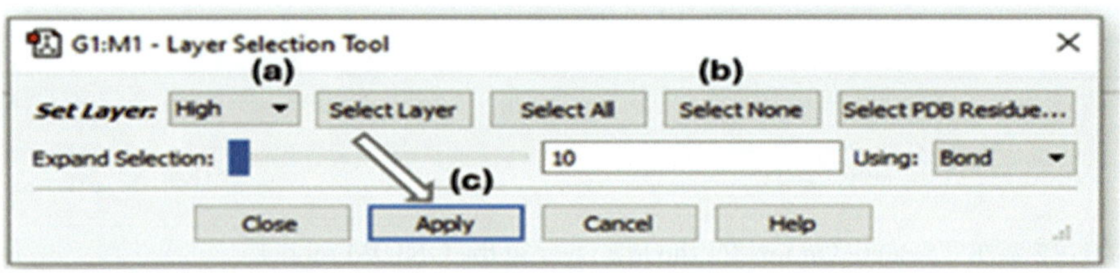

FIGURE 4.19 High layer selection tools tab.

Tube: Molecule appears as a tube model with no indication as to bond type. Atom types are indicated by color bands on the tubes.

Wireframe: Molecule appears as a series of thin lines, colored by atom type and indicating bond type via the number of lines.

None: Atoms are invisible (useful only with multilayer models) (Figure 4.21).

After changing the visualization for the low and high layers in the tube and ball & bond type, it will look as shown in Figure 4.22.

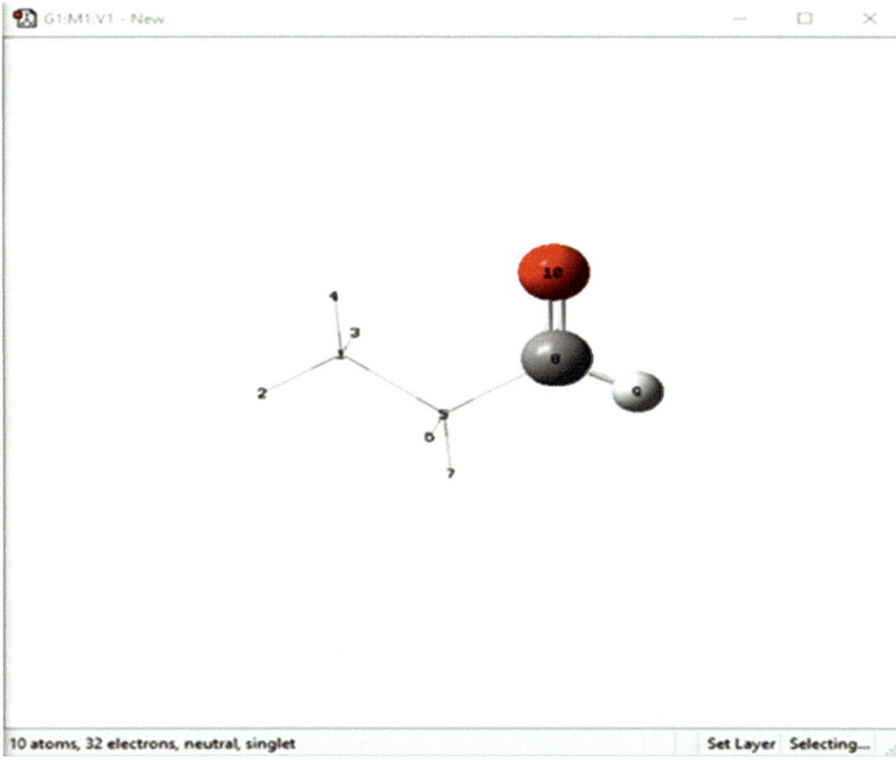

FIGURE 4.20 Selection of the high layer and then the respective change in the propanal molecule job.

Step 7: After setting up the layers, next, we will be setting up the calculation. We now must specify the type of calculation in each layer. For our present calculation, we may select the MP2/6–31G (d, p) for the high accuracy region and HF/3–21G or AM1 for the low accuracy calculation (Figures 4.23 and 4.24).

High layer Calculation Setup Dialog: ***MP2/6–31g (d, p)***

Low layer Calculation setup dialog: ***HF/3–21g***

Gaussian will specify the *link atom* automatically for this simple calculation. It is useful to look at your input file (propanal.com or propanal.gjf) before submitting the job. You will see that each atom is assigned to a level H(High) or L(Low) layer

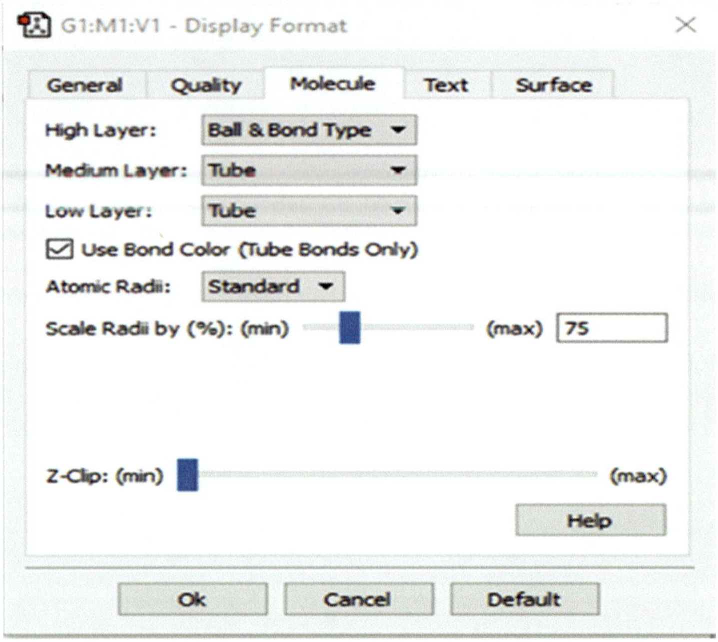

FIGURE 4.21 Molecule visualization display format.

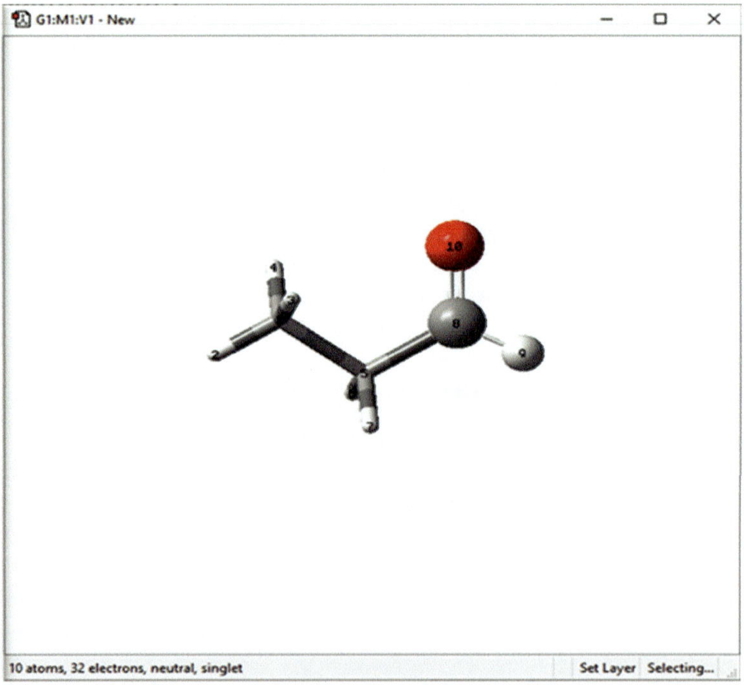

FIGURE 4.22 2-layer ONIOM model applied to propanal.

in the preview file. The carbon atom, which is the link between the low and high accuracy layers, has an additional "H 8" on the input line. This specifies that in the high-accuracy calculation, the atom is replaced by H, and it is bonded to atom 8 (the carbonyl carbon). Your numbering of the atoms may be different; it depends on your construction of the molecule. To see the atom numbering, use *view/labels*. This will show how the atoms are numbered in the input file. If the input file does not create the proper link atom, I think it is easiest to fix this in the input file directly. There are other ways to do this, which I think are more complicated.

Step 8: The preview of the Gaussian Calculation Setup shows all the details provided for the calculation. It's always a better idea to check the details before submitting the job. If everything looks good, then submit the job by saving an input file in your local directory. The keyword indicates the layer assignment for the atom, one of high, medium, and low; these may be abbreviated to their first initial letter (Figure 4.25).

Step 9: You can run the calculations and visualize frequencies as before. You will see that the carbonyl stretch does include some motion in your low accuracy region, and you can also compare the energy for different types of calculations performed so far (Figure 4.26).

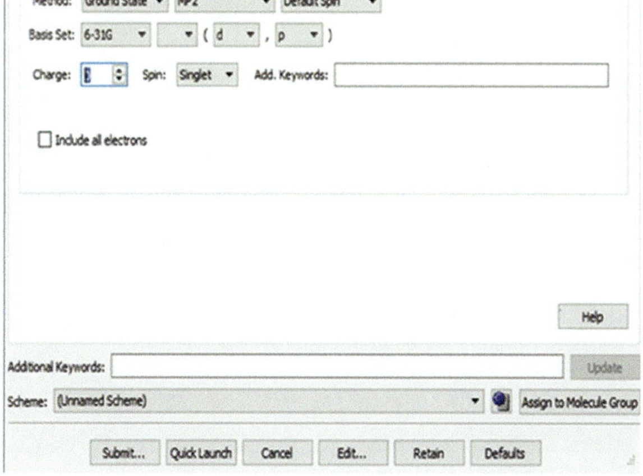

FIGURE 4.23 Gaussian calculation setup for high-layer method tab details.

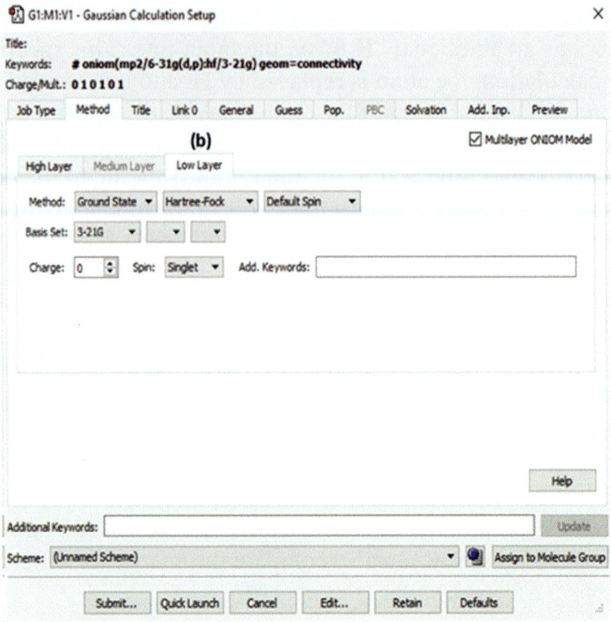

FIGURE 4.24 Gaussian Calculation Setup for low-layer method tab details.

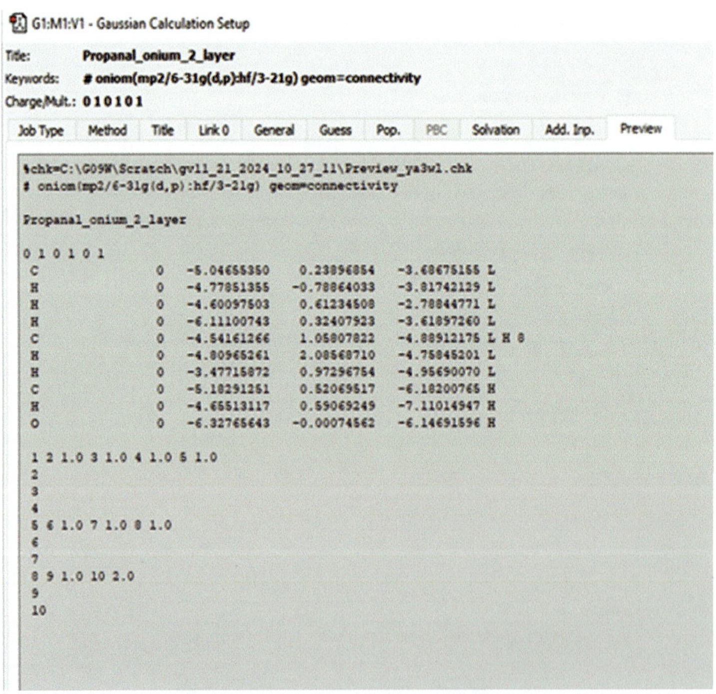

FIGURE 4.25 Gaussian calculation setup details in the preview tab.

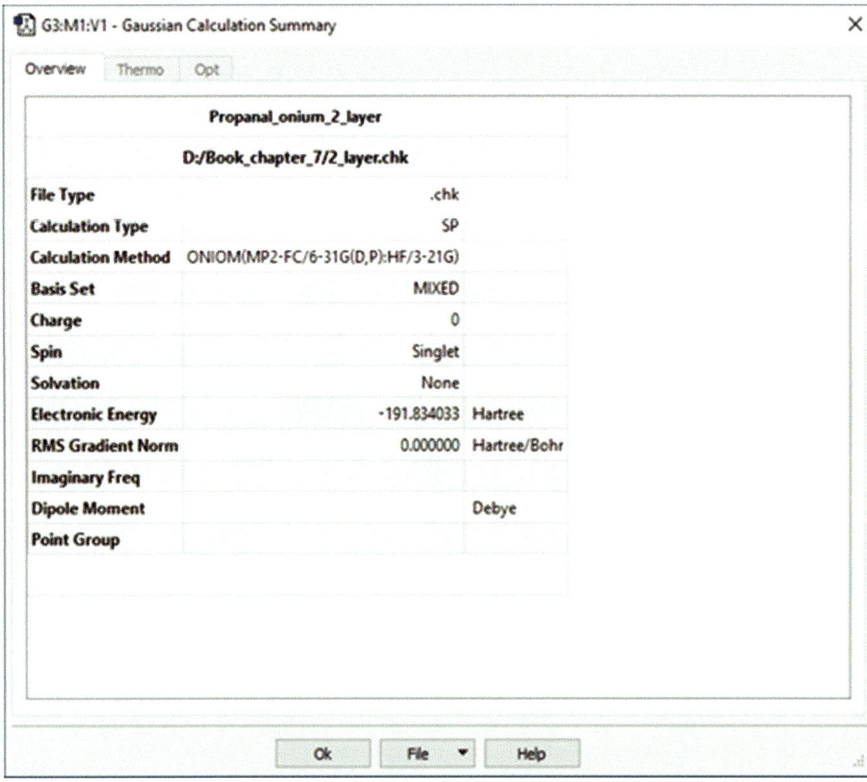

FIGURE 4.26 Gaussian calculation result summary for 2-layer calculations.

4.3.3 ONIOM Calculation Set-Up for Propanal Using the 3-Layer Model

Let us finally extend the model one more step and use a three-layer ONIOM model. This example is a little small for such a complicated approach, of course, but it is easy to envisage how such an approach can be useful. It is definitely advantageous to make gradual changes in the computational level.

Step 1: To select the medium layer—First, press *control and shift and select* **(a)** the CH_2 group of propanal molecule one by one and **(b)** then in the layer selection tool, open and select from the top-down menu *medium* and finally **(c)** press *apply.* The molecule will appear as illustrated in Figure 4.25. The representation used here is a medium layer as tube, low layer as wireframe, and high layer as ball and bond type. For details, please see Step 6 of Section 4.3.2 (Figures 4.27 and 4.28).

Step 2: Gaussian calculation setup for the three layers. We use formaldehyde in the high layer, using *MP2 / 6–31G (d, p),* then select the CH_2 group for the second, medium layer, and will do an *HF/6–31G (d, p)* calculation there,

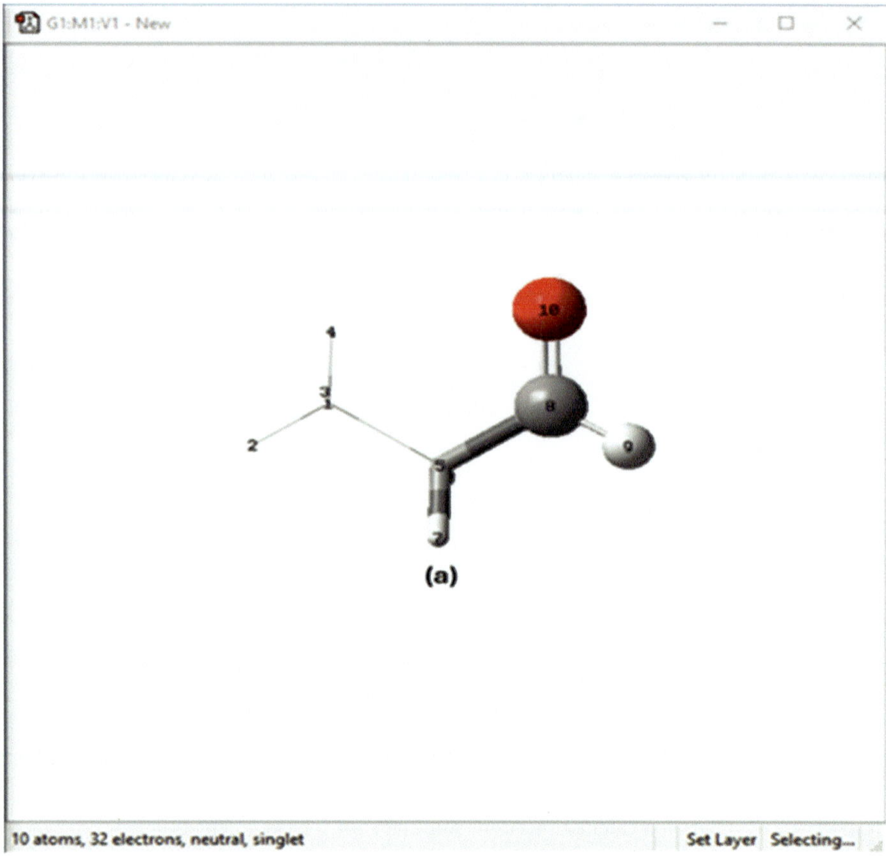

FIGURE 4.27 3-Layer propanal molecule for ONIOM calculation.

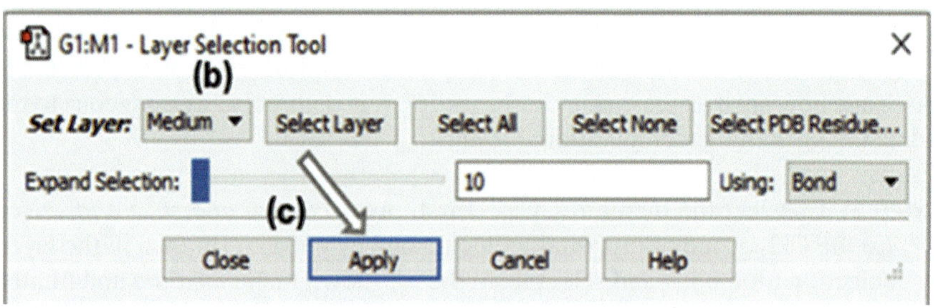

FIGURE 4.28 Selection of the medium layer for the ONIOM calculation.

while finally we will use *HF/3–21G* calculations for the final CH_3 group. Now check the method section of the Gaussian calculation setup and provide the details for each layer as follows (Figures 4.29–4.31).

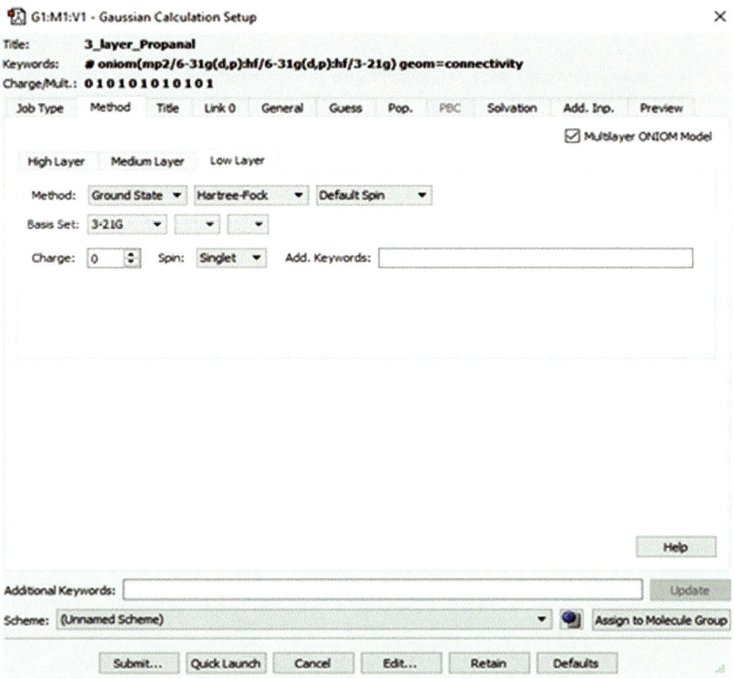

FIGURE 4.29 Method tab for the low-layer ONIOM calculation.

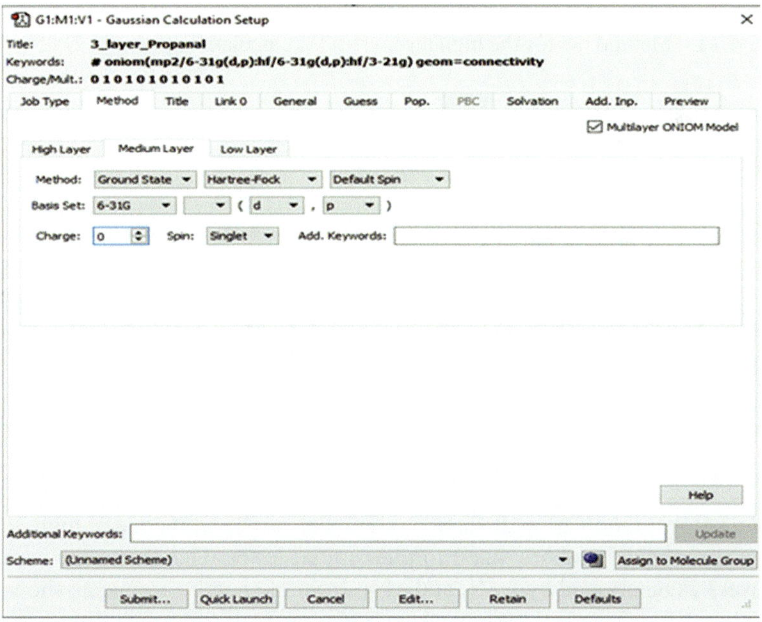

FIGURE 4.30 Method tab for the medium layer ONIOM calculation figure 4.31 method.

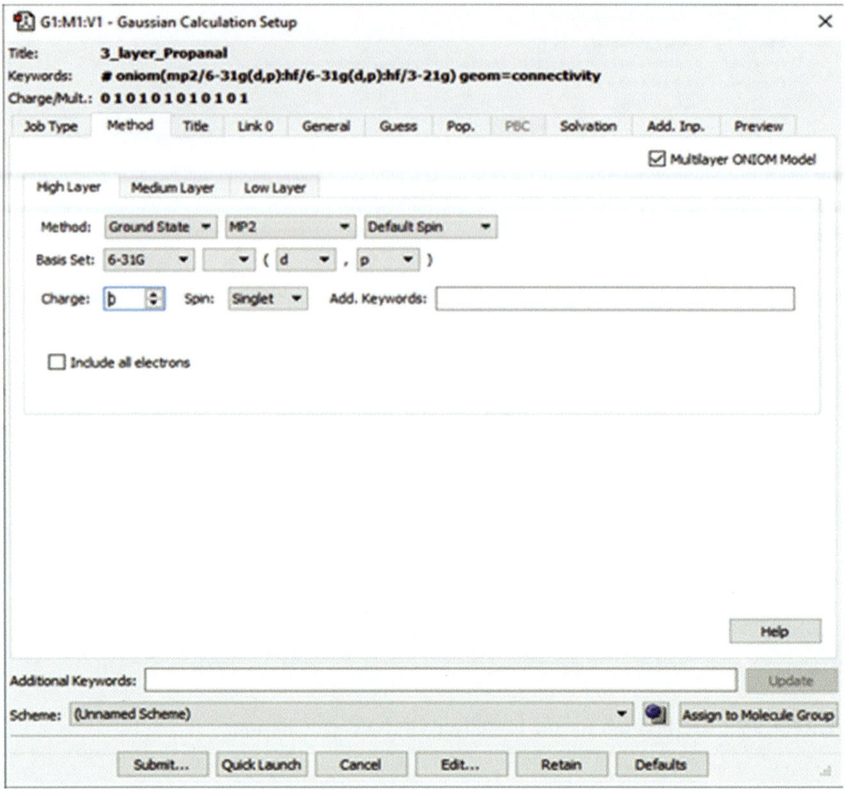

FIGURE 4.31 Method tab for the high layer ONIOM calculation.

a. **Low Layer:** *HF/3–21G*
b. **Medium Layer:** *HF/6–31G (d, p)*
c. **High Layer:** *MP2 / 6–31G (d, p)*

Step 3: After setting up the calculation for each layer, the next step will be to save and submit the job for the three-layer calculations. The panel given below presents a preview of the Gaussian Calculation Setup that GaussView 6.0 will create when a save or job submit operation is requested (Figure 4.32).

Job route:

 # oniom (mp2/6-31g(d,p):hf/6-31g(d,p):hf/3-21g)

Layer is a keyword indicating the layer assignment for the atom, one of high, medium, and low; these may be abbreviated to their initial letter. The atoms are assigned to three layers as mentioned by L, H, and M in front of specific atoms, as shown in the preview input file. Two link atoms are assigned as well for connecting low to high

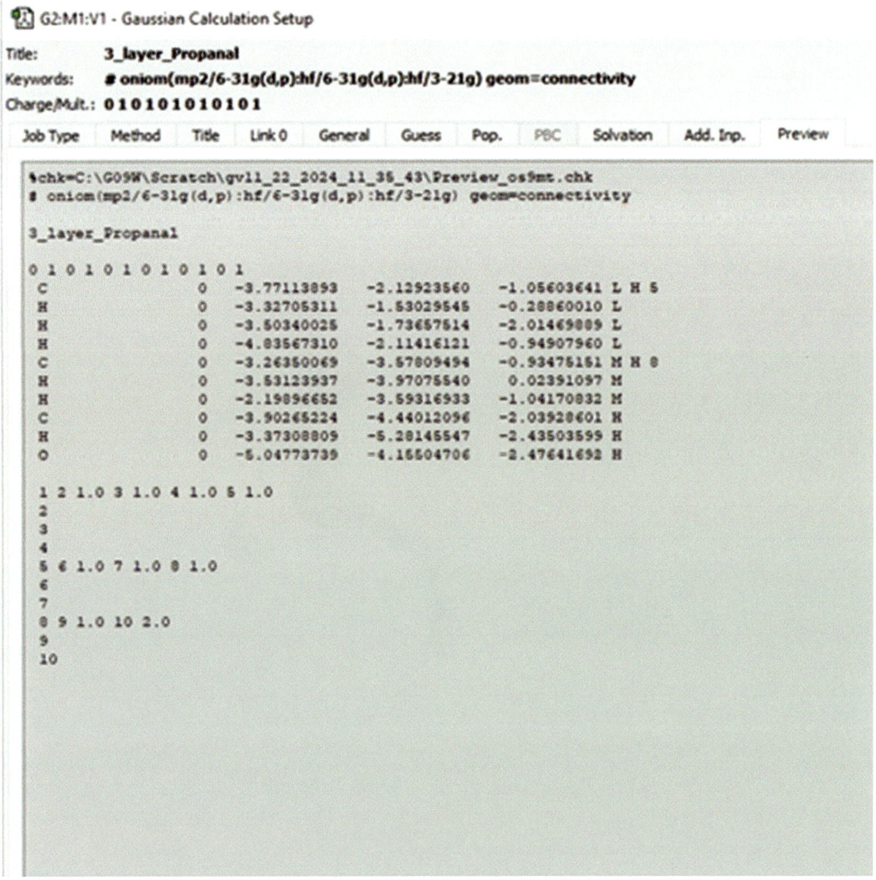

FIGURE 4.32 Preview for the three-layer ONIOM calculation.

by (L H 5) and medium to high by atom number (M H 8), as shown in the preview. Atoms number 5 and 8, as shown below in the figure, denote the link atoms between the different layers, such as 5 being a link atom between the low and high layers. Similarly, the atom number 8 denotes the link atoms between the medium and high layers, respectively (Figure 4.33).

Step 4: The Gaussian calculation summary shows results for the molecules. By comparing the energy for the calculation setup for two and three layers, you will see how much difference there is in the calculation setup. As propanal is a small molecule, as an exercise, you can optimize the entire molecule at a higher level, such as MP2, and compare all the sets of different energies and frequencies from the carbonyl bond stretch (Figure 4.34).

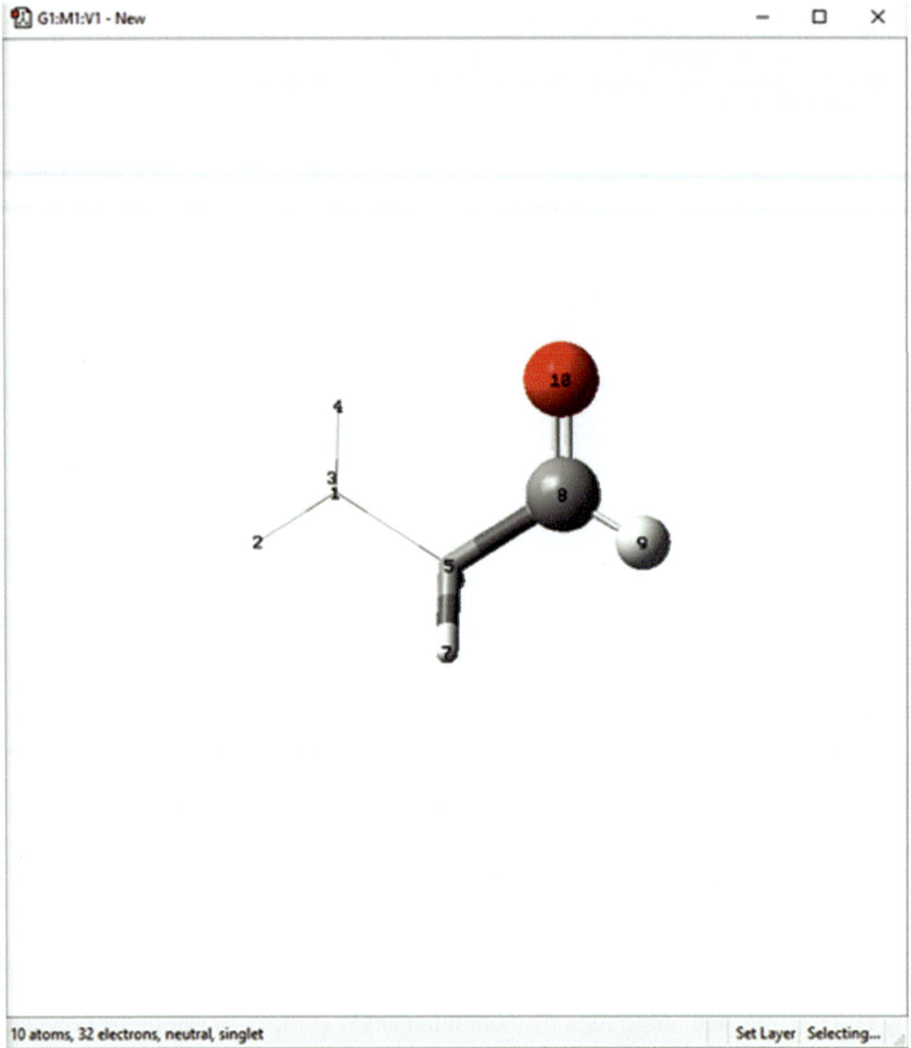

FIGURE 4.33 Three layers of propanal molecule for the ONIOM calculation.

4.3.4 ONIOM 3-Layer Calculation for a Complex Structure of Deoxyadenosine (DAn)

After understanding the ONIOM calculations by applying them to a small molecule such as propanal, now apply the three-layer calculations to a biological compound such as deoxyadenosine.

Step 1: Use the *biological fragment palette* to choose deoxyadenosine-free residue as shown in GaussView (Figure 4.35).

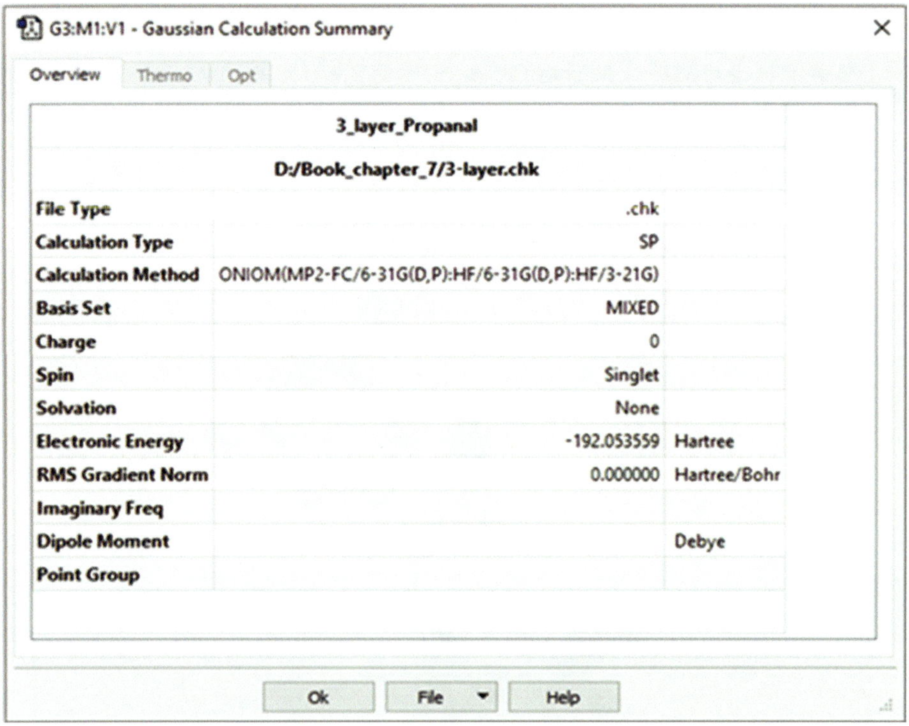

FIGURE 4.34 Gaussian calculation summary for the three-layer ONIOM calculation.

Step 2: Assign all atoms to the low layer. Choose select layer from the Tools menu and then click *Select All*. Verify that the current layer is set to *low*. The display will look like the following figure. Click *Apply*. Instead of manual selection of atoms one by one, you can also use the *expand selection* tool by selecting the central atom. Just expand the selection and adjust the atoms selected (Figure 4.36).

Step 3: Assign the following atoms to the medium layer: all atoms in five- and six-member rings. Use any selection method you want; it will usually take two or more operations to assign all these atoms. For example, you can select the central carbon atom in the ring and the Expand *Selection slider* to add more atoms. Adding too few atoms is usually easier than adding too many. Click Apply.

Step 4: Add additional atoms as necessary (use the center mouse button [or Shift+Click] to select, and then click *Apply*) until the selection is correct, as in the following figure. The atoms in the *medium* layers are now displayed as tubes (Figure 4.37).

Step 5: Assign the following atoms to the high layer: all oxygen and nitrogen functional groups. The wireframe displays the high layer, while the ball and bond type display the low layer. Click Close. The atom layer assignment is

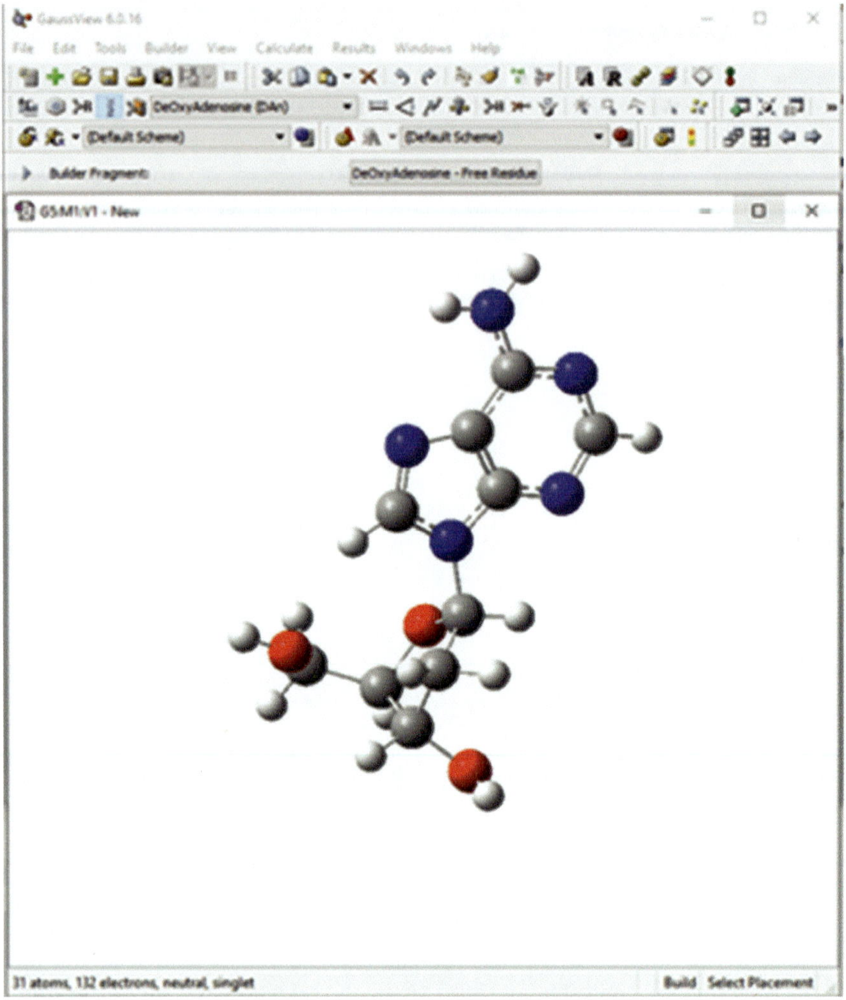

FIGURE 4.35 Deoxyadenosine free residue molecule.

now complete. After selecting the layer, the molecule will appear as shown in Figure 4.38.

Step 6: Setting up an *ONIOM* job. This example uses the molecule that was built and the layers that were defined in the final example in the Building Tutorial, and it continues from those steps.

6.1 Select the Gaussian calculation setup from the calculation menu.

6.2 On the method tab, *check Multilayer ONIOM Model*. Three tabs appear (high, medium, and low) in the *Method section of GaussView edit*,

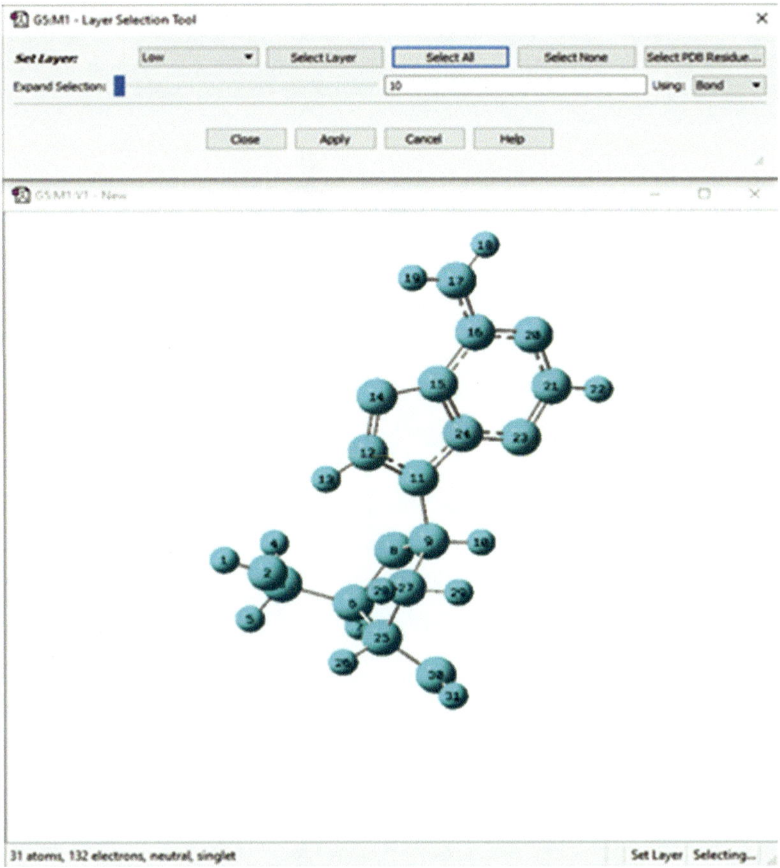

FIGURE 4.36 Deoxyadenosine free residue molecule after selecting all atoms as low layers.

which you can use to select the methods for treating each layer of atoms (Figure 4.39).

6.3 Click on the high tab: *MP2/6–311 +g(d,p)* job

6.4 Click on the medium tab, then select an *AM1* job (Figure 4.40)

6.5 Click on the low tab and specify a molecular mechanics calculation using the *Amber force field* (Figure 4.41)

6.6 These settings place the following line in the job's route section:

 # oniom(mp2/6-311+g(d,p):am1:amber) geom=connectivity

Step 7: You can now go on to fill out the rest of the dialog box details, and the job will be ready to be saved and run. Before submitting the job, you can check the preview of the job, and it will look like as shown in Figure 4.39.

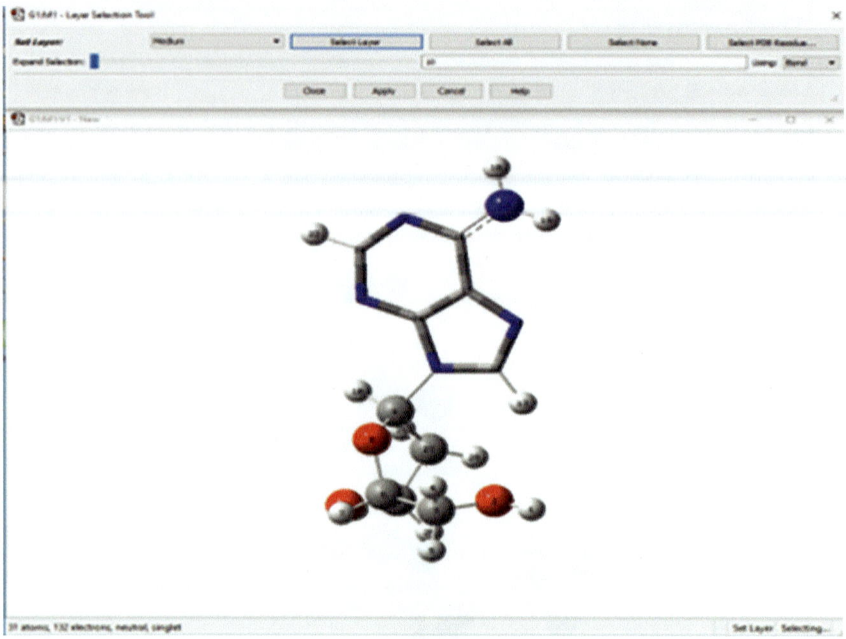

FIGURE 4.37 Deoxyadenosine free residue molecule after selecting atoms as low and medium layers.

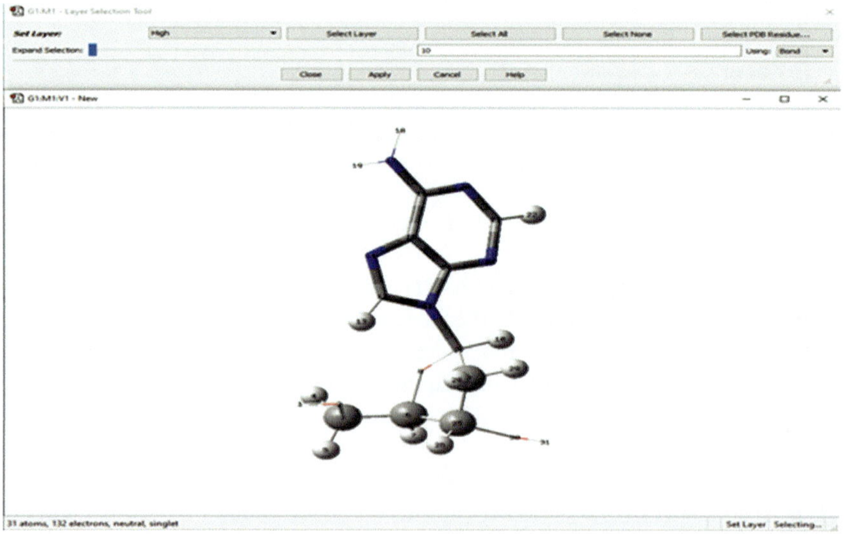

FIGURE 4.38 Deoxyadenosine free residue molecule after selecting atoms as low, medium, and high layer atoms.

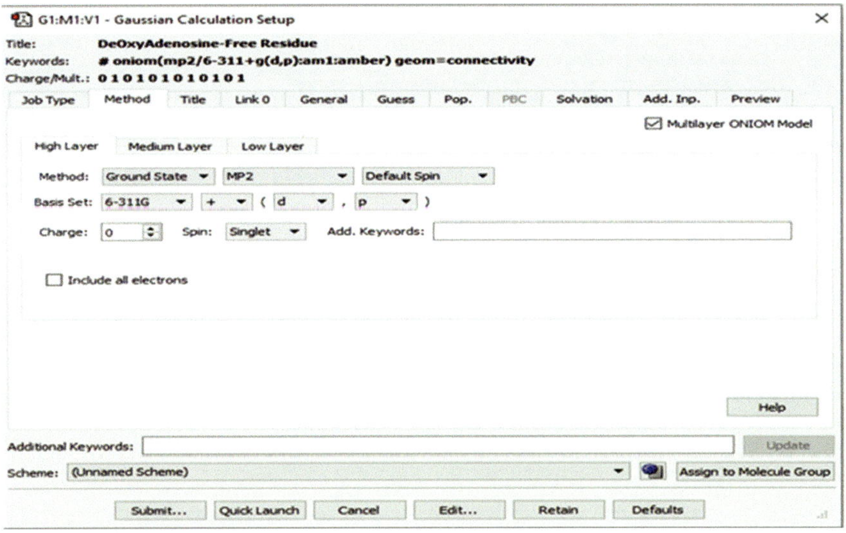

FIGURE 4.39 Deoxyadenosine free residue molecule details for the high layer in the method tab.

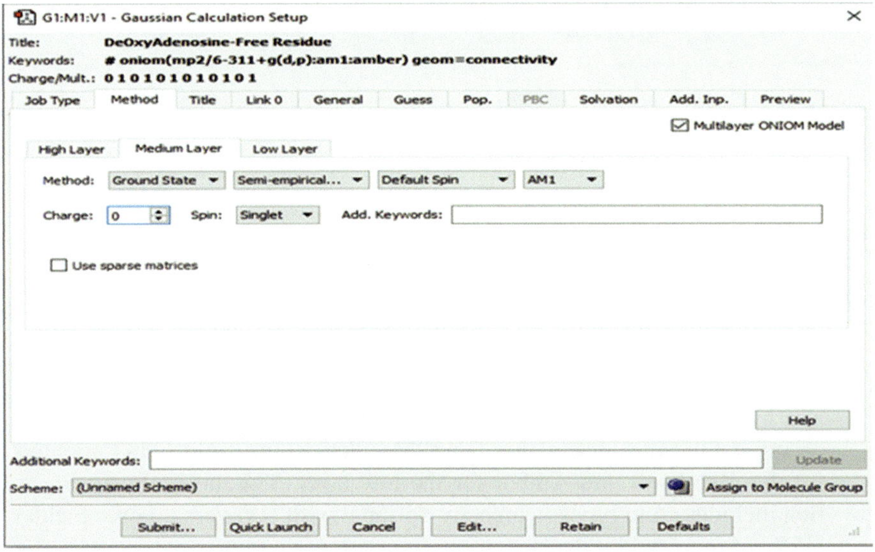

FIGURE 4.40 Deoxyadenosine free residue molecule details for the medium layer in the method tab.

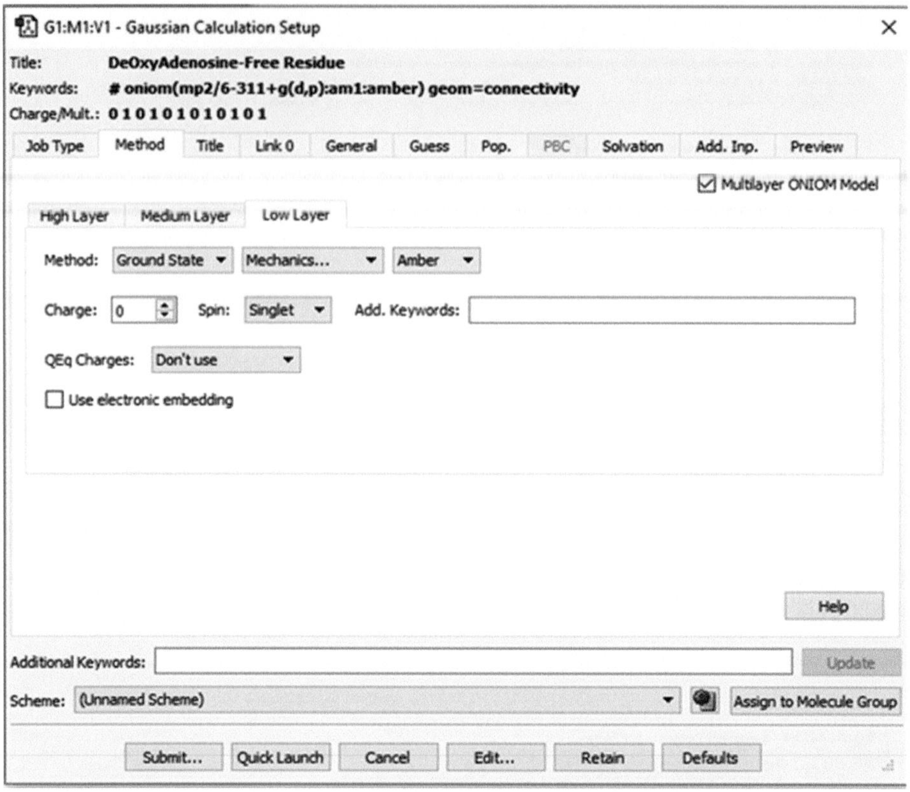

FIGURE 4.41 Deoxyadenosine free residue molecule details for the low layer in the method tab.

Step 8: Preview of the submitted job. Here is an input file created with GaussView that includes the additional input required for the Amber MM method. It also illustrates the use of per-layer charges and spin multiplicities (Figure 4.42).

Step 9: The initials of each layer, L, M, and H, represent the low, medium, and high layers assigned to respective atoms of the molecules. Here is an input file for a three-layer ONIOM calculation using a DFT method for the high layer and Amber for the low layer. The molecule specification includes atom types (which are optional with (Universal Force Fields) UFF but required by Amber). Keep in mind, atom types apply to both the primary atom specifications and the link atoms. We recommend starting with reference [4,5] below in order to take the next step in learning about ONIOM. It contains detailed information about performing MO:MM calculations.

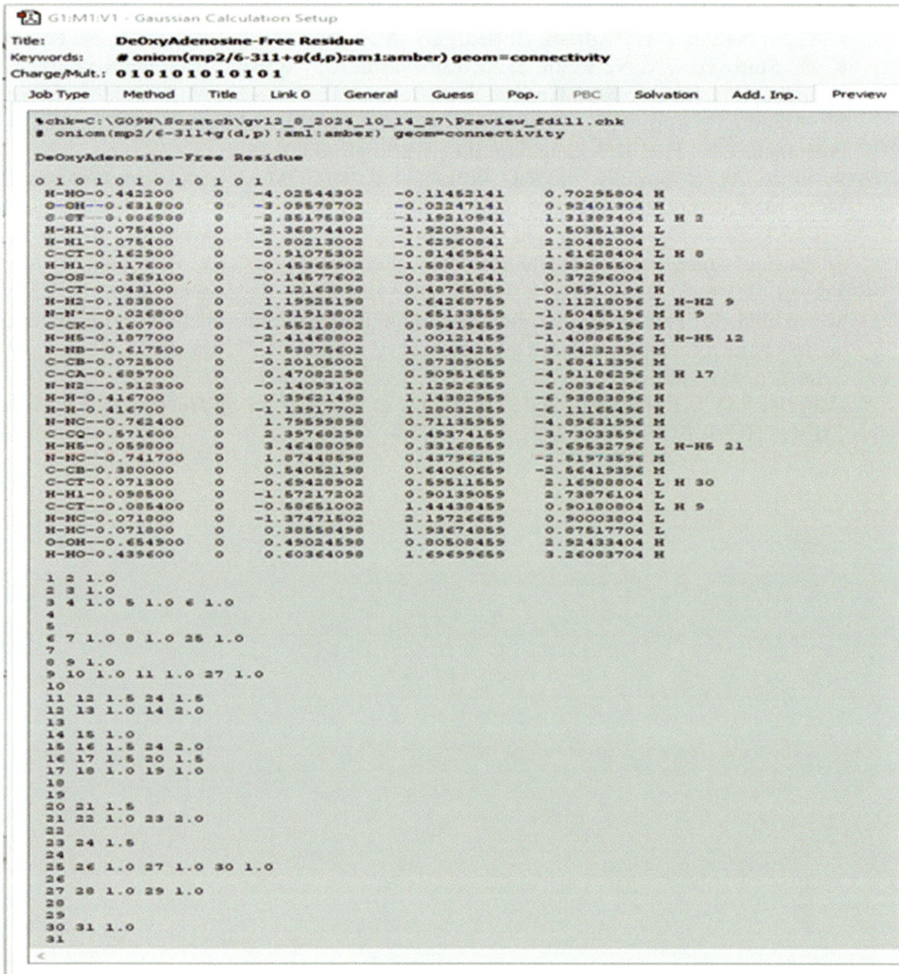

FIGURE 4.42 Deoxyadenosine free residue molecule preview details before submission.

REFERENCES

1. GaussView, Version 6, Dennington, Roy; Keith, Todd A.; Millam, John M. Semichem Inc., Shawnee Mission, KS, 2016.
2. Gaussian 09, Revision D.01, Frisch, M. J.; Trucks, G. W.; Schlegel, H. B.; Scuseria, G. E.; Robb, M. A.; Cheeseman, J. R.; Scalmani, G.; Barone, V.; Petersson, G. A.; Nakatsuji, H.; Li, X.; Caricato, M.; Marenich, A. V.; Bloino, J.; Janesko, B. G.; Gomperts, R.; Mennucci, B.; Hratchian, H. P.; Ortiz, J. V.; Izmaylov, A. F.; Sonnenberg, J. L.; Williams-Young, D.; Ding, F.; Lipparini, F.; Egidi, F.; Goings, J.; Peng, B.; Petrone, A.; Henderson, T.; Ranasinghe, D.; Zakrzewski, V. G.; Gao, J.; Rega, N.; Zheng, G.; Liang, W.; Hada, M.; Ehara, M.; Toyota, K.; Fukuda, R.; Hasegawa, J.; Ishida, M.;

Nakajima, T.; Honda, Y.; Kitao, O.; Nakai, H.; Vreven, T.; Throssell, K.; Montgomery, J. A., Jr.; Peralta, J. E.; Ogliaro, F.; Bearpark, M. J.; Heyd, J. J.; Brothers, E. N.; Kudin, K. N.; Staroverov, V. N.; Keith, T. A.; Kobayashi, R.; Normand, J.; Raghavachari, K.; Rendell, A. P.; Burant, J. C.; Iyengar, S. S.; Tomasi, J.; Cossi, M.; Millam, J. M.; Klene, M.; Adamo, C.; Cammi, R.; Ochterski, J. W.; Martin, R. L.; Morokuma, K.; Farkas, O.; Foresman, J. B.; Fox, D. J. Gaussian, Inc., Wallingford CT, 2016.

3. Svensson, M., S. Humbel, and R.D. Froese, et al., ONIOM: A multilayered integrated MO+MM method for geometry optimizations and single point energy predictions. A test for Diels–Alder reactions and Pt(P(t-Bu)3)2+H2 Oxidative Addition. *The Journal of Physical Chemistry*, 1996. **100**: p. 19357–19363.
4. Vreven, T., K.S. Byun, and I. Komáromi, Dapprich, S., Montgomery, J. A. Jr., Morokuma, K., Frisch, M. J., Combining quantum mechanics methods with molecular mechanics methods in ONIOM. *Journal of Chemical Theory and Computation,* 2006.**2**. p. 815–826.
5. Dapprich, S., I. Komáromi, and K.S. Byun, et al., *Journal of Molecular Structure:* THEOCHEM, 1999. **461**: p. 1–21.

5 Transition State Analysis

5.1 INTRODUCTION

Transition state (TS) analysis is a technique employed to examine the TS of a chemical reaction. The TS is a theoretical arrangement that exists between reactants and products in a chemical reaction. According to Gandour and Schowen, "Every dynamical problem in biochemistry is most elegantly and economically articulated through [TS] theory". The catalytic efficacy of enzymes originates from the interaction between enzyme and substrate structures in the TS; thus, comprehending this efficacy necessitates an understanding of these structures and interactions [1,2]. A number of constrained assumptions and the incomplete understanding of the actual form of the reacting system's potential surface certainly restrict the accuracy of the computation of an absolute reaction rate within the framework of transition state theory (TST) [3].

Computational chemical reaction modeling begins with the localization of stationary points along the reaction path on the molecular potential energy surface (PES). These points are located by geometry optimization methods. As an example, the reactant and product complexes serve as the final destinations of a chemical reaction. These structures represent the lowest locations on the surface of molecular potential energy. Structures that have additional minima along the reaction path are (meta) stable reactive intermediates. The TSs are the maximum points along the chemical reaction path, which are also the saddle points on the surface of molecular potential energy. We can estimate many thermodynamic and kinetic properties [4], such as rate constants (from the height of reaction barriers, using TST) and equilibrium constants (from the relative depth of the minima), once we have determined these stable structures (minima) and the TSs connecting them (Figure 5.1).

Experiments on reactions involving multiple steps often only reveal the reactant and the end product. In such instances, understanding the reaction mechanism—including all the intermediates and TS structures—can be done by calculating the minimal energy path of the reaction. More efficient and less harmful synthetic processes can result from using this information to build catalysts.

5.2 POTENTIAL ENERGY SURFACE

Taking into account all potential structures for a particular chemical formula is the initial step in bringing the theory closer to the experiment. They provide a comprehensive description of the PES for any particular chemical formula. The PES has $3N-6$ coordinate dimensions, where N is the number of atoms ≥ 3, and is described as a hypersurface by the potential energy of a set of atoms across all conceivable atomic configurations. The three-dimensionality of Cartesian space is the source of this dimensionality. So, every structure, which is a point on the PES, may be

DOI: 10.1201/9781032694733-5

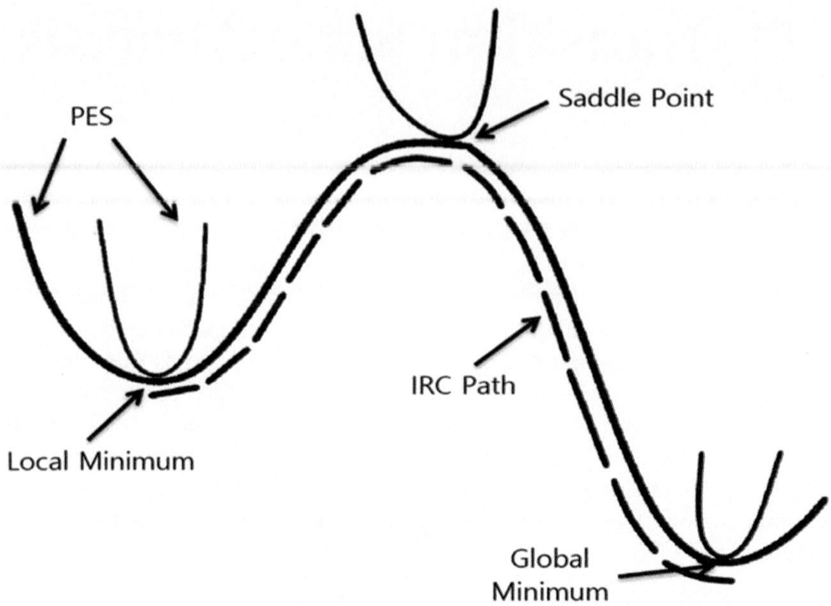

FIGURE 5.1 Energy levels for intermediates, transition states, and products are indicated on a Potential Energy Surface (PES) diagram that shows energy changes throughout a reaction route.

described by a vector X given by the Cartesian coordinates of atom i, x_i, y_i, and z_i and where X $(x_1, y_1, z_1, x_2, y_2, z_2, \ldots, x_N, y_N, z_N)$. Since it includes an arbitrary origin, this X expression does not provide a unique definition of the structure.

We can decrease the dimensionality without changing the structure by eliminating the three dimensions related to translation in the x, y, and z directions, such as by making the molecular center of mass be at the origin and by requiring the principal moments of inertia to align in increasing order along the x, y, and z axes. In this manner, we can remove the three dimensions related to rotation about those axes.

With the sole specification of atom "a" nature in frame I, no degrees of freedom are present. The length of the bond is the only variable in frame II, which has one degree of freedom. Two more degrees of freedom, one for each bond length or valence angle, are needed to determine where atom "c" is located in frame III. Multiple approaches to pinpointing atom "d" exact location are shown in Frame IV. It should be noted that all of them requires the addition of three new degrees of freedom, expressed in either Cartesian or internal coordinates.

One other approach to understanding the reduced dimensionality is to picture building an atomic structure vector (Figure 5.2). In this scenario, it would be easiest to picture the PES dimensions as internal coordinates, such as bond lengths and valence angles. Therefore, there are no degrees of geometric freedom involved in choosing the initial atom, as the atom itself specifies the origin.

The second atom's location is defined by how far away the first one is. Accordingly, the bond length is the only degree of freedom in a two-atom system, which should

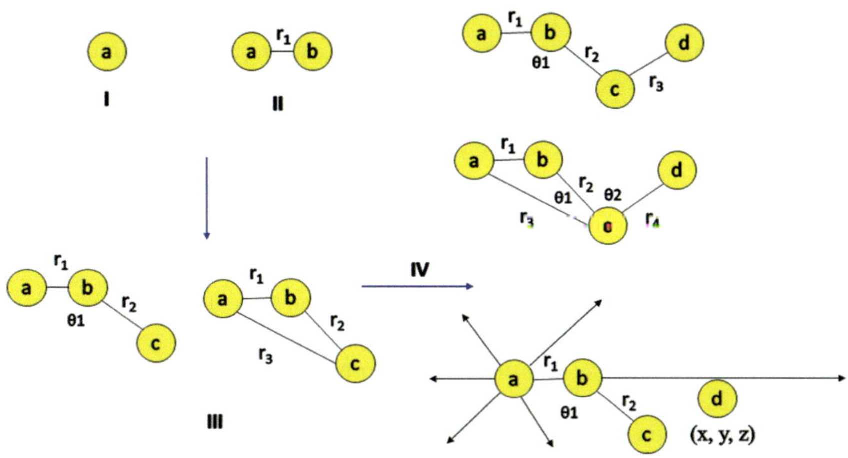

FIGURE 5.2 A single atom is shown in Frame I, two bond lengths are shown in Frame II, three valence angles are added in Frame III, and four degrees of freedom are added in Frame IV. These frames illustrate the methods used to define molecular structures.

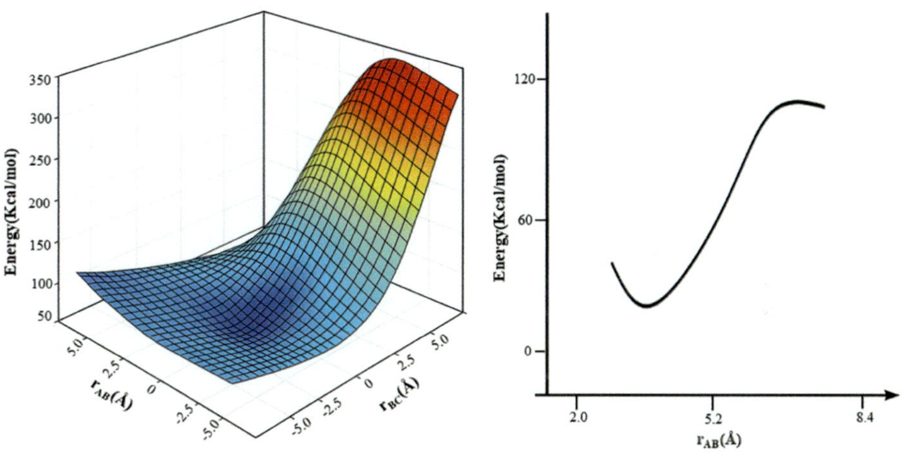

FIGURE 5.3 An illustration of a hyperslice of the PES with energy as a function of two bond lengths in three dimensions.

be equivalent to $3N-5$ degrees of freedom in a linear molecule. Either the distances to the two atoms that came before it or the angle between the two bonds that were defined to a common atom up to this point can be used to identify the third atom. There should be three degrees of freedom in the three-atom system if collinearity is not enforced. The location of each extra atom can be described using three coordinates. The description of those coordinates can be envisioned in a number of ways. They can take the form of a bond length plus two valence angles, or a bond length plus a valence angle and a dihedral angle, as shown in Figure 5.3.

The complete PES for the hypothetical molecule ABC requires four dimensions to represent all three coordinate degrees of freedom and one energy dimension. The three-dimensional plot presented illustrates a hyperslice of the complete PES. It depicts energy as a function of the bond lengths AB and BC, while maintaining a constant angle ABC, specifically the value corresponding to the global minimum on the entire PES. In the section beneath this surface, the energy is plotted as a function of the length of the AB bond, while the length of the BC bond has been held constant at its equilibrium value corresponding to the global minimum.

Local minima on PESs, which represent optimal molecule structures, and saddle points, which have no slope in either direction. It can have a downward curvature for one coordinate and an upward curvature for all the others. As the lowest energy barriers on paths connecting minima, saddle points can be connected to the chemical idea of a TS, according to simple calculus. Therefore, for any set of atoms, a full PES details every conceivable chemical structure and the isomerization pathways that link them.

Due to the high dimensionality involved, visualizing entire PESs for polyatomic molecules presents significant challenges. Figure 5.3 illustrates the pertinent energy curves or surfaces in reduced dimensionality. Typically, PESs are analyzed by slicing through them, focusing on one or two coordinates, such as a bond length. When the slice is defined as a hyperplane, it indicates that all non-visualized coordinates are assigned fixed values. The non-visualized coordinates assume values that minimize potential energy, contingent upon the specified value of the visualized coordinate(s). In this scenario, the curve represents the minimum energy path corresponding to changes in the visualized coordinate within a single visualized dimension. Consequently, numerous structures exhibiting identical values for a specific internal coordinate can establish a genuine continuous pathway connecting any two structures on a PES.

Using the full PES or the part of the PES that would be chemically accessible in the experimental system being modeled, along with the principles of statistical mechanics, one can calculate ensemble averages of physical observables and estimate equilibrium populations in cases with multiple stable molecular structures.

5.3 INTRINSIC RATE COORDINATE

When it comes to understanding and forecasting the kinetics of chemical reactions, PES may be indispensable in two ways. To comprehend the dynamical behavior of a complicated chemical system, one must first dissect the system into its basic processes. From a theoretical perspective, one can qualitatively evaluate the probable significance of different processes by looking at the PESs of hypothetical reactions. Reactions with low barriers are more likely to play a significant role, whereas those with high barriers are less likely to do so. In addition, the PES is useful for determining the range of any elemental reaction. There are two possible ways in which a bimolecular condensation can take place: either all at once with one expected TS structure or in stages with two TS structures. This type of condensation forms two new bonds between the reacting species. Beginning with a condensation and ending with a unimolecular rearrangement, the two basic reactions form the basis of the sequential process.

To claim that a process as a whole incorporates two distinct TS structures, however, is to assume that the reacting system plots out a trajectory on the PES. Most of the time, when chemists consider a system's motion on a PES, they picture a certain path known as the minimum energy path, or occasionally the intrinsic reaction coordinate (IRC). As shown in Figure 5.1, the maximum energy potential (MEP) is the path that a ball rolling on a surface would follow if its velocity were endlessly damped at every point as it descends from a saddle point to a minimum. In mass-weighted coordinates, the PES is also known as the MEP, which is the path that follows the steepest gradient at every point.

As a conceptual tool, the MEP and its relationship to TS structure(s) are very helpful. It should be noted that the free energy of populations of molecules, not the potential energy of individual molecules, governs the equilibria and dynamics of reacting systems. If the free energy reflects the temperature distribution of the system's entities, then the system can be seen as a cloud passing over the PES, with Boltzmann's statistics showing that the cloud's density decreases as it rises. The net distribution is still determined by temperature, even though individual molecules in the cloud may be exchanging energy to rise and fall relative to the PES.

One way to look at a reacting system is as a cloud passing over the PES on its way to a mountain pass, with the TS structure serving as the saddle point. Nevertheless, the cloud's passage across the pass is not required to occur directly over the TS structure. Cloud height and pass width determine how far to the left and right of the TS structure cloud particles can pass at will. When the pass is extremely small, for instance, we say that the reaction is experiencing an entropic bottleneck; this means that, apart from the reaction coordinate, there is very little room for variation in the other degrees of freedom. This means that although the TS structure can tell us something about the pass height and local topology

(thanks to the Taylor expansion of the surface around the stationary point), it is still just one out of many molecules that make the transition from reactants to products. Therefore, the maximum possible rate of molecular transit from reactants to products over a given time interval is the ratio of the two directions.

Till now, we have not shown any connection between a theoretical quantity linked to the individual elementary reactions and their forward and reverse rate constants, even though theory is useful for breaking down complex systems into their component series of reactions. Reactions involving high-energy TS structures are inherently expected to move at a slower rate than low-energy TS structures, but for a more rigorous quantitative analysis, we need to use more complex models that describe the connection between the activated complex's properties and kinetics. In this category, TST needs to be discussed.

5.4 TRANSITION STATE THEORY

TST assumes that the activated complex represents a population of molecules in equilibrium with one another and also in equilibrium with the reactant, A. This partitions the molecules between an irreversible forward reaction and further it deactivates back to A, with a (reverse) rate constant. As per the Maxwell–Boltzmann

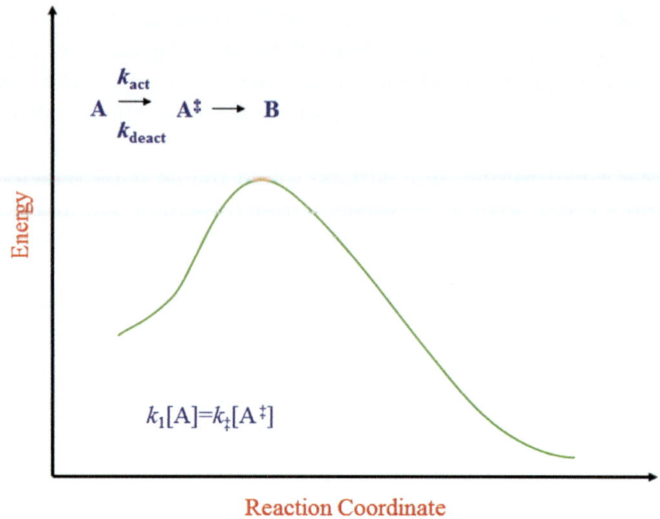

FIGURE 5.4 An equilibrium constant between activated complexes and reactants is highlighted in a graph that shows the rate constant connection for a bimolecular process.

statistics, it relates to the fact that it relies on the number of molecular collisions with energies close to activation energy. It thus points out that the molecules in the TS are in equilibrium with the reactant, and the molecules reaching the TS always react irreversibly (Figure 5.4).

$$k_1 = \frac{k_{\ddagger}\left[A^{\ddagger}\right]}{[A]} = k_{\ddagger}K^{\ddagger} \tag{5.1}$$

Here, $K^{\ddagger}$ is the equilibrium constant between the activated complex and the reactants. Using the usual relationship between free energy and the equilibrium constant, we may write

$$K^{\ddagger} = e^{\frac{-(G_{\ddagger}-G_A)}{k_BT}} \tag{5.2}$$

where the difference in free energy between the activated complex and the reactants is referred to as the activation free energy. Using the thermodynamic concepts, we write

$$G = U_0 + PV + k_BT \ \ln Q \tag{5.3}$$

Here, Q is the partition function.

$$K^{\ddagger} = e^{\frac{-\left[\left(U_{\ddagger,0}-PV_{\ddagger}-k_BT\ \ln Q_{\ddagger}\right)-\left(U_{A,0}-PV_A-k_BT\ \ln Q_A\right)\right]}{k_BT}} \tag{5.4}$$

$$= \frac{Q_{\ddagger}}{Q_A} e^{\frac{-(U_{\ddagger,0}-U_{A,0})}{k_B T}} e^{\frac{-(PV_{\ddagger}-PV_A)}{k_B T}} \tag{5.5}$$

$$\approx \frac{Q_{\ddagger}}{Q_A} e^{\frac{-(U_{\ddagger,0}-U_{A,0})}{k_B T}} \tag{5.6}$$

Assuming that *PV* changes are negligible in activation of *A* leads to

$$k_1 = k_{\ddagger} \frac{Q_{\ddagger}}{Q_A} e^{\frac{-(U_{\ddagger,0}-U_{A,0})}{k_B T}} \tag{5.7}$$

In the exponential term, the difference in internal energies between A and $A_{\ddagger}$, which includes the zero-point energy, can be easily calculated from the electronic struc ture calculation and the frequency calculation of the ZPVE for the minimum energy and TS structures corresponding to A and $A_{\ddagger}$, respectively. Furthermore, Q_A may be quickly calculated due to the availability of A frequencies. However, the nature of $Q_{\ddagger}$ needs to be defined here.

The TS structure needs to be identified as a species with a minimum of $3N-7$ degrees of freedom, with the reaction coordinate representing the "missing" degree of freedom. The electronic, translational, and rotational parts of the partition function related to the transition structure can therefore be defined in the standard manner. Our goal is to disentangle the partition function for the reaction coordinate degree of freedom with respect to the vibrational component.

$$k_1 = \frac{k_{\ddagger}}{1-e^{\frac{-h\omega_{\ddagger}}{k_B T}}} \frac{Q_{\ddagger}}{Q_A} e^{\frac{-(U_{\ddagger,0}-U_{A,0})}{k_B T}} \tag{5.8}$$

Here $Q_{\ddagger}$ is the reduced partition function over the $3N-7$ bound degrees of freedom and $\omega_{\ddagger}$ is the "vibrational frequency" associated with the reaction coordinate. Using a power series expansion for the exponential function of $\omega_{\ddagger}$, we truncate after the first two terms.

$$k_1 = \frac{k_{\ddagger}}{1-\left(1-\frac{-h\omega_{\ddagger}}{k_B T}\right)} \frac{Q_{\ddagger}}{Q_A} e^{\frac{-(U_{\ddagger,0}-U_{A,0})}{k_B T}} \tag{5.9}$$

$$k_1 = \frac{k_{\ddagger} k_B T}{h\omega_{\ddagger}} \frac{Q_{\ddagger}}{Q_A} e^{\frac{-(U_{\ddagger,0}-U_{A,0})}{k_B T}} \tag{5.10}$$

It is the real inverse time constant linked with motion along the reaction coordinate, not the imaginary frequency derived from normal harmonic oscillator analysis.

This becomes the correct way to think about the vibrational frequency in this context. However, it is exactly motion along the reaction coordinate that converts the activated complex into product B, hence both the vibrational frequency and the real inverse time constant are equal:

$$k_1 = \frac{k_B T}{h}\frac{Q_{\ddagger}}{Q_A}e^{\frac{-\left(U_{\ddagger,0}-U_{A,0}\right)}{k_B T}} \tag{5.11}$$

For a bimolecular reaction, this reaction takes the form:

$$k_1 = \frac{k_B T}{h}\frac{Q_{\ddagger}}{Q_A Q_B}e^{\frac{-\left(U_{\ddagger,0}-U_{A,0}-U_{B,0}\right)}{k_B T}} \tag{5.12}$$

5.5 IMPLEMENTATION IN GAUSSIAN

A quasi-Newton or eigenvector-following algorithm is used to execute the optimization in the Synchronous Transit-Guided Quasi-Newton (STQN) Method, which initially employs a linear synchronous transit or quadratic synchronous transit technique to approach the quadratic region around the TS. By default, it optimizes using redundant internal coordinates when minimization is concerned. Utilizing an empirical estimation of the Hessian and appropriate initial structures, this approach will efficiently converge to the real transition structure. In contrast to other approaches, STQN only requires the structures of the reactants and products to be entered, rather than a hypothesis for the transition structure.

The *Opt* keyword is used to request this approach together with the *QST2* and *QST3* options. As an input, *QST2* needs the reactant and product molecular specifications, in that order, but *QST3* needs all three, in that order: the reactant, the product, and an initial structure for the TS. All molecular requirements require the same atomic order. Please be aware that neither *QST2* nor *QST3* should have the TS option selected. An input file that can be used to find the transition structure for the reaction $SiH_4 \rightarrow SiH_2 + H_2$ can be seen on the left side of the screen. After the reactant's title and molecular specifications, the product follow (Figure 5.5).

The above file creates a checkpoint file named sih2_h2_qst2.chk. Gaussian then uses this to store wavefunction and geometry information for later restarts or frequency jobs.

For #P RHF/6-31G(d) Opt = (QST2, AddRedundant, CalcFC) SCF = Tight

- **#P** → "Print" keyword: gives extra verbose output.
- **RHF/6-31G(d)**: This uses the Restricted Hartree–Fock method with the 6-31G(d) basis set.
- **Opt=(QST2, AddRedundant, CalcFC)**: This performs geometry optimization using the **QST2 method**. This requires a transition state search requiring two structures: reactants and products. *AddRedundant* tells

```
%chk=sih2_h2_qst2.chk
#P RHF/6-31G(d) Opt=(QST2,AddRedundant,CalcFC) SCF=Tight

SiH2 + H2 → SiH4 (QST2 endpoints) — Reactant complex (This is the title section)

0 1                                  #Charge and Multiplicity (all electrons are paired)

Si     -1.93457936   0.10280362   0.00000000
X       1  1.0
H      -1.44459640  -1.28313170   0.00000000
H      -1.44457084   0.79576224   1.20024997
H      -1.44457084   0.79576224  -1.20024997
H      -3.40457936   0.10282169   0.00000000

A1=120.0
rHH=0.66
```

FIGURE 5.5 An illustration of the input file structure used by the GAUSSIAN software to calculate transition states, which includes reactant and product molecular characteristics. The description above is for the reactant which is then followed by product in Figure 5.6.

Credit: Gaussian User Manual (https://gaussian.com/qst2/)

Gaussian to automatically add internal coordinates beyond the minimum Z-matrix set. *CalcFC* requests an initial Hessian calculation to improve convergence.

- **SCF = Tight** Tightens the self-consistent field convergence criteria.

Here we define SiH_2 (three hydrogens attached to Si at 1.48 Å). This will not be visible in the input file as cartesian coordinates are generally pulled out.

Thereafter we added an approaching H_2 molecule and attaching the second with bond length rHH. The angle A1 controls the geometry of the approach. Let us see this in steps as below:

Si

The first atom is silicon, at the origin by default. The first atom always defines the coordinate system.

X	**1**	**1.0**

A dummy atom X placed 1.0 Å away from Si. Dummy atoms (X) do not contribute to electrons; they are used to define bond angles and dihedrals in Z-matrices.

H	**-1.44459640**	**-1.28313170**	**0.00000000**

The first hydrogen atom is bound to atom 1 (Si) at 1.444 Å. The angle between H–Si–X is 55°.

H	**1.44457084**	**0.79576224**	**1.20024997**

Second hydrogen bound to Si at 1.48 Å with angle H–Si–X as 55°.The dihedral angle with the previous hydrogen H is 180°, placing this H opposite to the first one.

H	**2**	**A1**	**3**	**90.0**

The third hydrogen is bound to Si at a distance 1.48 Å. The angle H–Si–X = A1 (a variable angle). The dihedral angle is defined as relative to previous atoms is equal to 90°. The hydrogen is one of the atoms of the approaching H_2 molecule.

H	**5**	**rHH**	**1**	**180.0**	**2**	**0.0**

The fourth hydrogen or the other atom of H_2. molecule. This is bound to atom 5 (the previous H) at distance **rHH** (the H–H bond length). The angle with Si = 180.0° and torsion = 0.0°, placing it colinear with the previous H to make a diatomic H_2 molecule.

5.5.1 Variable Definitions

The Si–H forming bond is set long at 1.48 Å so as to mimic a reactant complex.

A1=120.0

The approach angle for the incoming H relative to Si and the dummy atom is 80°.

rHH=0.74

The H–H bond length fixed at 0.74 Å. A typical equilibrium bond length for H_2.

5.5.2 Analysis

Here, the **QST2** method will optimize the TS geometry between the reactants and products. The key outputs include (a) **Energy values** (look for `SCF Done:` for ground-state energy), (b) **Frequency calculations** (to confirm the TS has one and only one imaginary frequency), and (c) **Optimized geometries** with adjusted bond lengths, angles, and dihedrals. Here, for the reaction, $SiH_2 + H_2 \rightarrow SiH_4$ reaction, we will obtain the optimized geometry for both reactants and products. We shall also obtain the calculated energy barrier for the reaction. Here, we want to make sure that the optimized structure printing includes the H-H bond length—which are specifically interested. This is done by specifying the internal coordinate that binds those two atoms to the *Add Redundant* option. By default, the optimized structure printout will also contain the Si-H bond lengths. The three components of the input file for *Opt=QST3* will be the same as before: reactants, products, and an initial transition structure guess. The sections also pertain to molecular specifications. The output format for the optimized structure identified by *QST2* or *QST3* is identical to that of other geometry optimization types (Figure 5.6):

Figure 5.6 specifies SiH_4 product geometry in tetrahedral arrangement. All the Si–H bonds are equal (1.48 Å) and all H–Si–H angles are close to the tetrahedral angle (109.47°). The dihedral values (120°, –120°, 0°) ensure the four hydrogens are distributed symmetrically around Si. Together with our earlier reactant complex section (Figure 5.5), this completes the QST2 job input. Now Gaussian will interpolate between these two endpoints and search for the transition state connecting them.

SiH2 + H2 → SiH4 (QST2 endpoints) — Product (SiH4)

0 1

Si	-1.97734 0.13411 -0.05422
X	1 1.0
H	-0.67094 -0.94533 1.41705
H	-0.67093 -0.42572 1.71705
H	-1.36559 0.80243 -1.2118
H	-3.4408 0.06485 0.06577

FIGURE 5.6 A screenshot of the GAUSSIAN interface continuing from Figure 5.5 displaying the product species for QST2 method's transition state optimization choices.

Credit: Gaussian User Manual (https://gaussian.com/qst2/)

Si
Central silicon atom at the origin.

X	**1**	**1.0**

Dummy atom (X) 1.0 Å away from Si. This is again used to define bond angles and torsions. Does not represent a real atom.

H	**1**	**1.48**	**2**	**109.47**

First hydrogen attached to Si (atom 1) with Si–H bond length = 1.48 Å. The angle H–Si–X = 109.47° relates to a tetrahedral structure.

H	**1**	**1.48**	**2**	**109.47**	**3**	**120.0**

The second hydrogen, also 1.48 Å from Si. The angle H–Si–X is 109.47°. The dihedral angle relative to the first H = 120°, so the two hydrogens are spaced apart.

H	**1**	**1.48**	**2**	**109.47**	**3**	**120.0**

The third hydrogen, same bond length and angle but with dihedral angle a –120°, placing it opposite to the second H, creating tetrahedral distribution.

H	**1**	**1.48**	**2**	**109.47**	**3**	**0.0**

The fourth hydrogen, with the same bond length and angle but with dihedral angle of 0.0°, places this H orthogonal to the other three, completing the tetrahedral SiH_4 geometry.

Here, Atom 4 is one of the hydrogens and Atom 5 is another hydrogen. The purpose of *Redundant Coordinates* in Gaussian's QST2 method (for TS optimization) is to constrain specific bond distances or angles and to improve convergence

```
! Optimized Parameters !
    ! (Angstroms and Degrees) !
-----------------------------------------------------------------
! Name Definition   Value   Reactant Product  Deriv Info.  !
-----------------------------------------------------------------
! R1  R(1,2)        1.5204  2.2442  1.47  -DE/DX =  -0.0001  !
! R2  R(1,3)        1.6384  2.271   1.47  -DE/DX =   0.0     !
! R3  R(1,4)        1.4729  1.47    1.47  -DE/DX =   0.0     !
!...
-----------------------------------------------------------------
```

FIGURE 5.7 Optimized molecular structure output in GAUSSIAN, highlighting the bond lengths and geometries of the transition state.

during optimization. It is also used to guide the calculation to locate the correct TS by enforcing critical structural features. The 4 5 directive ensures that Gaussian explicitly considers this bond as a key structural feature when optimizing the TS. This helps Gaussian locate the correct energy pathway more effectively (Figure 5.7).

Figure 5.7 shows the excerpt of the "Optimized Parameters" table from the Gaussian transition state (or reaction path) job. It summarizes the key internal coordinates (bond lengths, bond angles, torsions) that Gaussian has been optimizing/ interpolating between the reactant and product geometries. Going row wise:

! R1	**R (1,2)**	**1.5204**	**2.2442**	**1.47**	**DE/DX = 0.0001 !**

Here R1 is the label for this parameter. R (1,2) is the bond length between atoms 1 and 2. The value of 1.5204 Å indicates an optimized bond length at the current geometry (e.g., transition state). The value of reactant i.e. 2.2442 Å indicates that the bond length is our starting reactant structure. The product at 1.47 is the bond length in our product structure. -DE/DX = -0.0001 shows that the derivative of the energy with respect to this coordinate; near zero means geometry is well-optimized in that coordinate.

! R2	**R (1,3)**	**1.6384**	**2.271**	**1.47**	**DE/DX = 0.0 !**

R2 (1,3) is the second bond parameter. R(1,3) is the bond between atoms 1 and 3 where the optimized value is 1.6384 Å (at TS or current point) which is in between reactant = 2.271 Å and product = 1.47 Å. Again -DE/DX = -0 indicates that derivative very small, showing this coordinate is converged.

! R3	**R (1,4)**	**1.4729**	**1.47**	**1.47**	**DE/DX = 0.0 !**

R3 the bond between atoms 1 and 4. The value is 1.4729 Å which sits between reactant (1.47 Å) and product (1.47 Å). This is expected in a transition structure where many coordinates interpolate between endpoints.

For a successful QST2 calculation, we shall find:

1. The optimized geometry for the transition state.
2. The **Activation Energy (ΔE‡)**—calculated as the energy difference between the TS and the reactants.
3. The **Reaction Energy (ΔE)**—calculated as the difference between products and reactants.

In any chemical reaction, the involvement of at least one TS is essential for the interconversion of stable geometries into alternative forms. Considering that many reactions occur via several elementary steps, it is crucial to take into account the TSs linked to each of these steps. If one of these steps exhibits a markedly slower rate compared to the others, leading to the overall reaction rate being predominantly governed by the conversion rate of the species involved in this particular step, it is identified as the rate-determining step. As a result, characterizing the TS for this step is a primary goal in elucidating the reaction pathway (mechanism). The TS that possesses the highest energy is commonly designated as the rate-determining TS. Instances exist where multiple TSs, possessing nearly equivalent energy levels, can each play a role in influencing the overall reaction rate. In multistep reactions, there must be at least one stable electronic geometry present between the reactants and products, which is identified as a reaction intermediate. Two categories of intermediates are present before specific rate-determining TSs, along with two categories of TSs, namely energy barriers, related to the formation of these intermediates [3].

Intermediate:

1. **Low Energy:** close in energy to reactants.
2. **High Energy:** significantly higher in energy than reactants.

Transition states:

1. **Low Barrier:** significantly lower in energy than the rate-determining TS.
2. **High Barrier:** close in energy to the rate-determining TS.

This chapter focuses on the chemical reaction pathways. The initial calculation method is the same as discussed in the previous chapter. In this chapter, a step-by-step method has been followed to calculate the reaction pathway of a reaction, calculating the energy changes at each step and understanding the TS to clarify how the reactant transfers into the product. The analysis is often done by studying the PES. Here are the basic steps to calculate the reaction pathway of a chemical reaction.

Step 1: Generating electronic structure of reactant and product molecules.
Step 2: Optimization of reactant and product molecules with a suitable basis set and finding the lowest energy state.
Step 3: Identification of TS using method (TS Berny, QST2, and QST3) to obtain the TS having only one imaginary frequency.
Step 4: IRC calculation of the calculated TS structure to get the reaction path.

Apart from this, some additional keywords are required to perform the calculation to collect the required data. Optimization and frequency calculations for the reactants and products will forecast the thermally corrected energy and enthalpy.

The procedure may be replicated for an alternative reaction pathway by initiating from another saddle point. This improves the investigation of supplementary potential transitions from reactants to products. This enables a comprehensive examination of the PES.

Let's take one example to better understand QST2 and IRC calculation.

Example 5.1

Conversion of CH_3O to CH_2OH. Here, the reactant is CH_3O, and the product is CH_2OH. We need to calculate the reaction pathway from reactant to product.

Using HF, usually with the 3–21G basis set, optimize the geometries of the reactant (CH_3O) and product (H_2CO + H radical) before calculating the TS for the intramolecular hydrogen transfer from the carbon atom of CH_3O to its oxygen atom using Gaussian [5]. After frequency analysis (all real frequencies) confirms that these geometries are local minima, estimate the TS structure, perhaps by interpolating between the reactant and product geometries or by manually moving the hydrogen atom in a visualization program like GaussView [6]. To carry out the TS optimization, use Gaussian's Opt=(TS,CalcFC,NoEigenTest) keyword. Make sure that the force constants are computed first (CalcFC) to direct convergence and prevent the early rejection of structures (NoEigenTest). After optimization, do a frequency calculation (Freq) to verify the nature of the TS and make sure that the hydrogen transfer vibration is represented by exactly one imaginary frequency. To confirm that the TS connects to the reactants and products smoothly, do an IRC calculation once the TS has been confirmed. To calculate the reaction pathway in both forward and backward directions, use the term IRC=(CalcFC, MaxPoints=50). To guarantee a smooth energy descent to the reactant and product wells, examine the IRC results. For a good understanding of the reaction process, it is advised to visualize the pathway using software such as GaussView. The successful completion of these procedures validates the validity of the TS structure for the hydrogen transfer process and offers a thorough understanding of the reaction pathway. The followings are the steps to calculate the reaction pathway.

Step 1: Optimization of each molecule (reactant and product) at a small basis set HF/3–21G. The basis set can be higher as per the requirement (Figure 5.8).

Step 2: After completing the optimization of both reactant and product molecules, shown in Figure 5.9, the next step is to copy the optimized reactant molecule and paste in the product window terminal, as shown in Figure 5.10

Now, again in the calculation setup, choose the optimization method and select the optimize to TS(QST2) (Figure 5.11) and run the calculation, as shown in Figure 5.12.

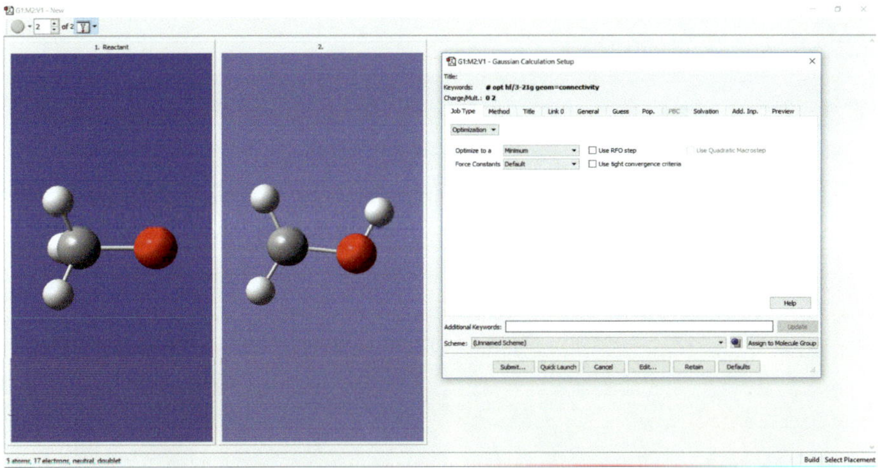

FIGURE 5.8 CH_3O and CH_2OH reactant initial molecule optimization visualization utilizing a small basis set.

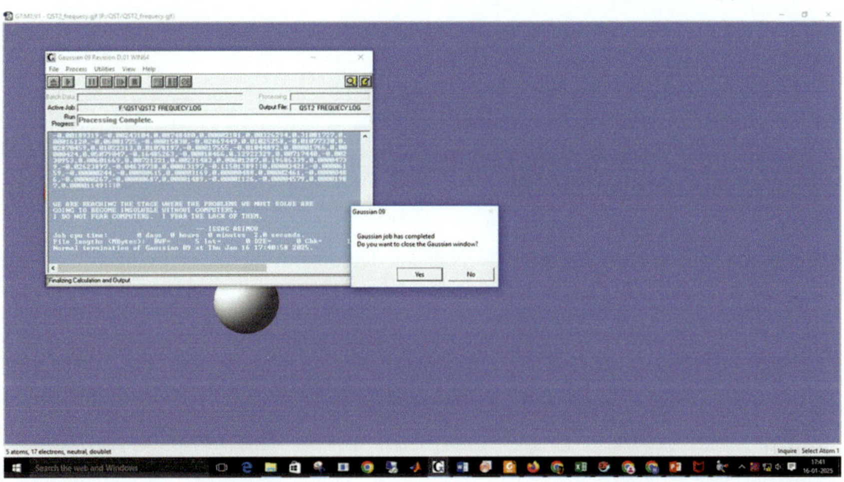

FIGURE 5.9 The structures of the optimized molecules of the reactant (CH_3O) and product (CH_2OH) are shown side by side following the initial optimization.

After successfully running the calculation, the H atom dissociates from the CH_3O molecule, as shown in Figure 5.13.

Now again go to the calculation terminal and do the frequency calculation (Figure 5.14).

Run the calculation keeping the basis set and methods the same (Figure 5.15).

After successfully completing the calculation, select result in the terminal (Figure 5.16) and open the frequency table.

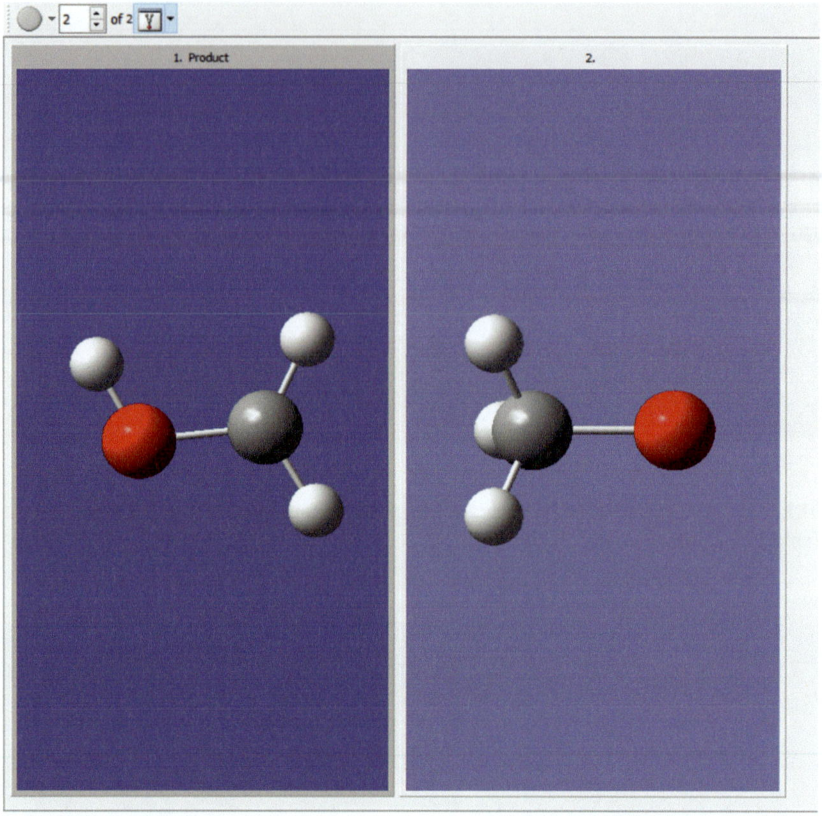

FIGURE 5.10 TS calculation setup screenshot demonstrating the copying of optimized reactant molecule data into the product window terminal.

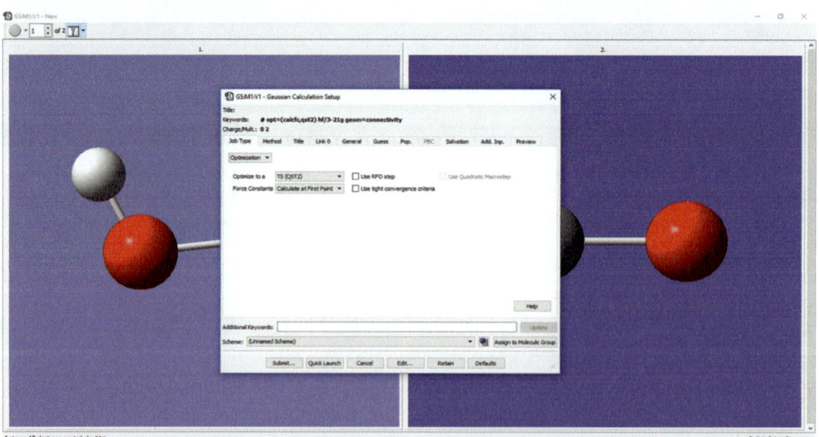

FIGURE 5.11 GAUSSIAN setup screen highlighting the selection of QST2 for transition state optimization.

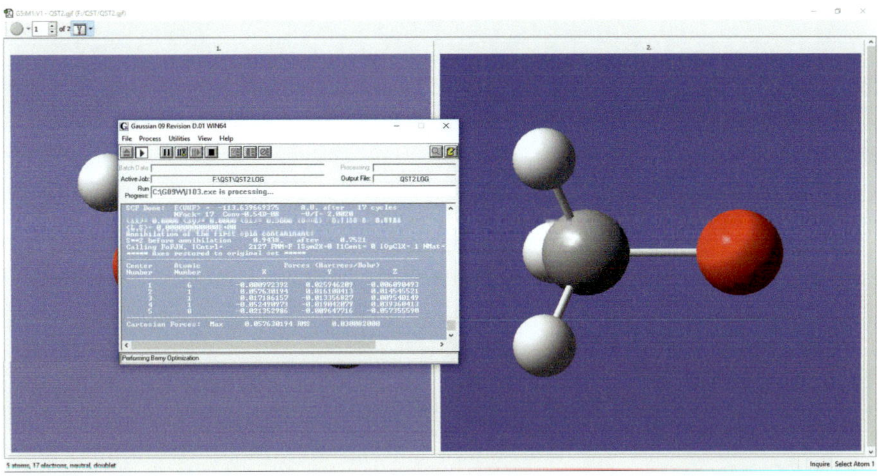

FIGURE 5.12 A screenshot of the GAUSSIAN software conductingthe QST2 method calculation for transition state optimization.

FIGURE 5.13 Illustration of the hydrogen atom's separation from the CH_3O molecule throughout the transition state.

As shown in Figure 5.17, there is one negative frequency (–2556.72). In IRC calculation, there should be one negative frequency. The reaction pathway between reactants and products is defined by the mode that corresponds to the negative frequency.

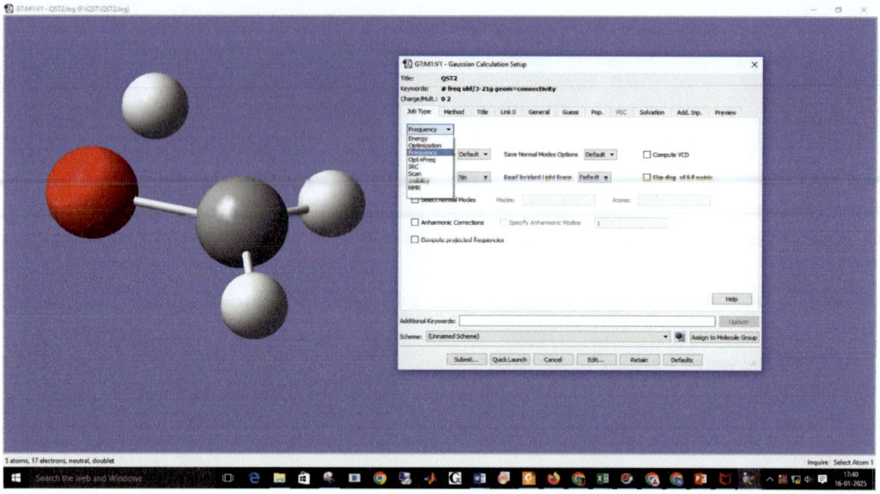

FIGURE 5.14 Screenshot the GAUSSIAN interface for frequency calculation using the same methodology and basis set as earlier stages.

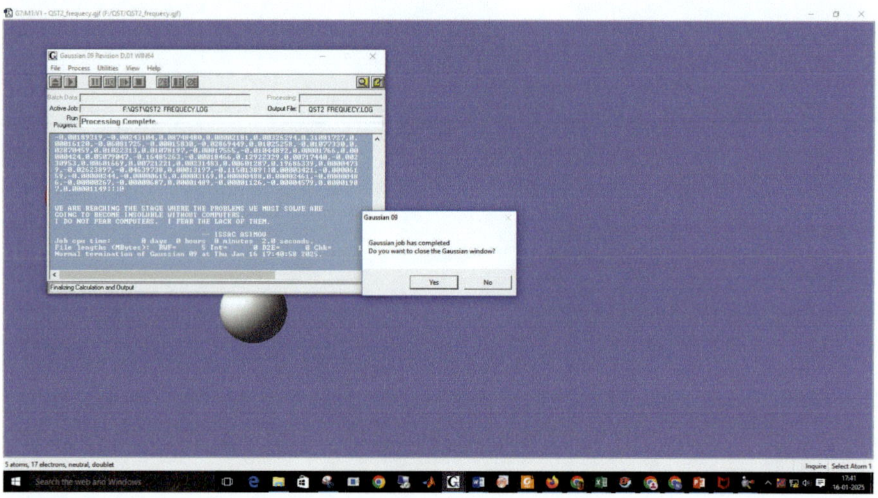

FIGURE 5.15 The GAUSSIAN surface displays the ongoing frequency calculating procedure.

This mode's visualization ensures that the TS produces the right reactants and products. A single negative frequency is necessary in TS calculations to verify that the structure is a TS, to make sure it connects reactants and products, and to allow accurate kinetic modeling. In reaction mechanism investigations, the computation becomes meaningless without it [7,8].

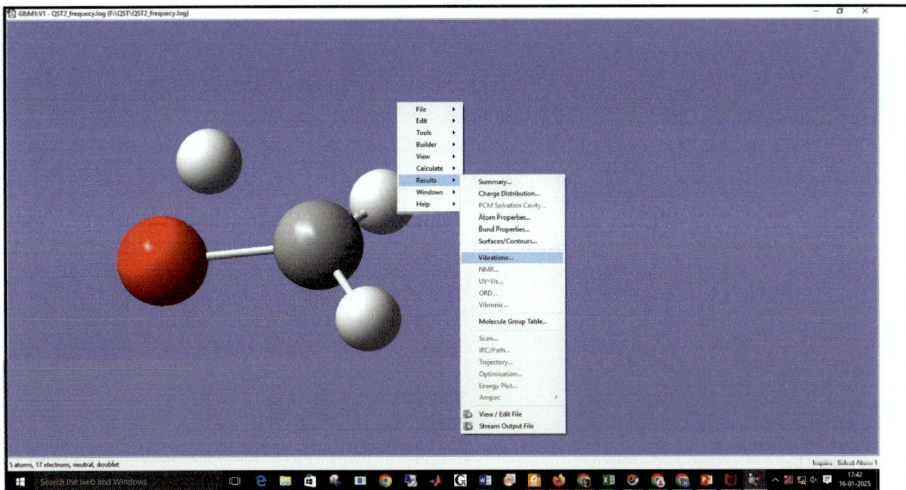

FIGURE 5.16 The frequency table can be opened from the GAUSSIAN terminal that displays the results of the calculation.

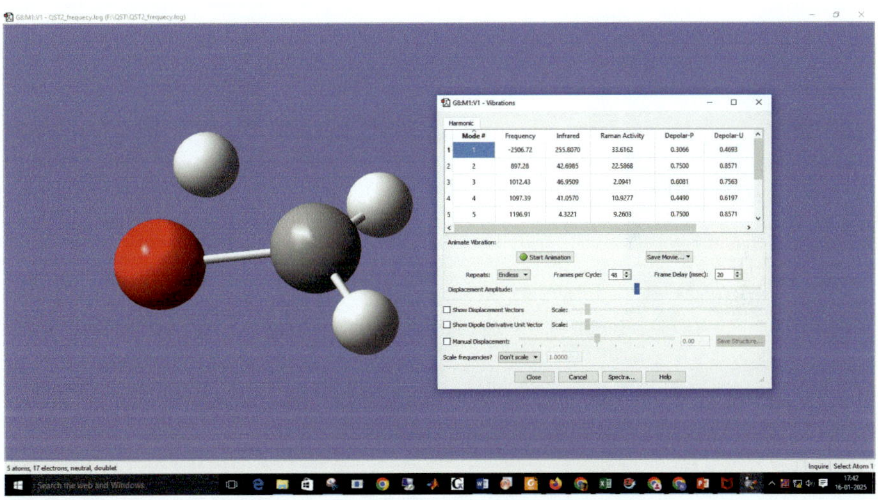

FIGURE 5.17 One negative frequency (-2556.72) is included in the frequency table to verify the existence of a legitimate transition state structure.

We can start animation by clicking the animation button, where we can see the movement of H atom between carbon and oxygen.

To make sure the TS connects the reactants and products, do an IRC calculation once the TS has been confirmed. IRC calculation should be performed on the confirmed TS structure. IRC calculation setup is shown in Figures 5.18 and 5.19.

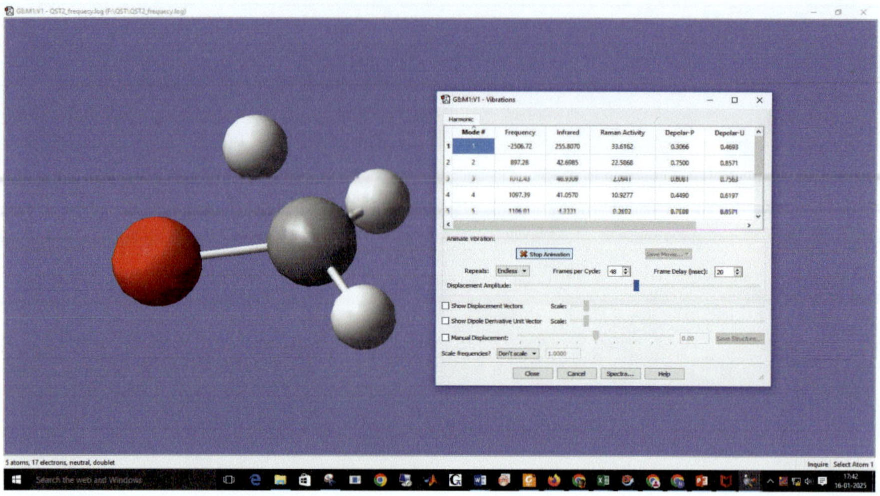

FIGURE 5.18 The GAUSSIAN interface to calculate the Intrinsic Reaction Coordinate (IRC) from the verified transition state.

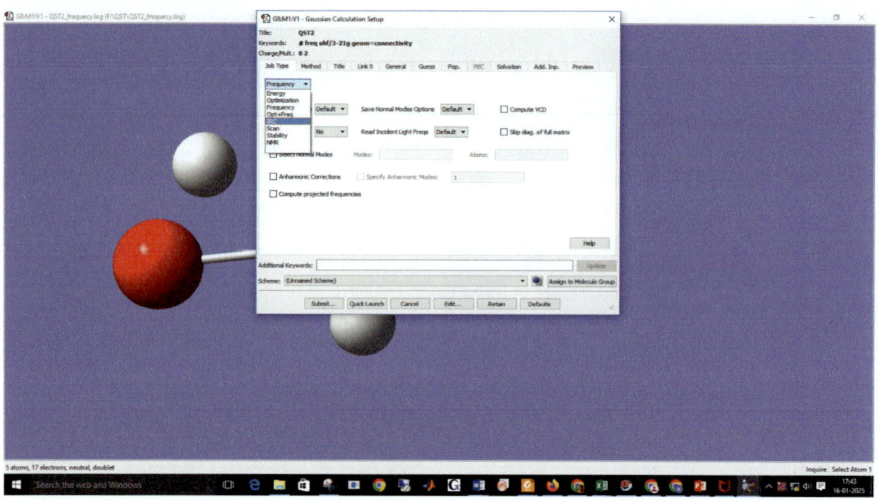

FIGURE 5.19 GAUSSIAN interface showing the IRC calculation process in progress.

Run the calculation keeping the method the same (shown in Figure 5.20).

After successfully completing the optimization, open the result section and go to the IRC option (Figure 5.21)

The required IRC path is shown in Figures 5.22 and 5.23.

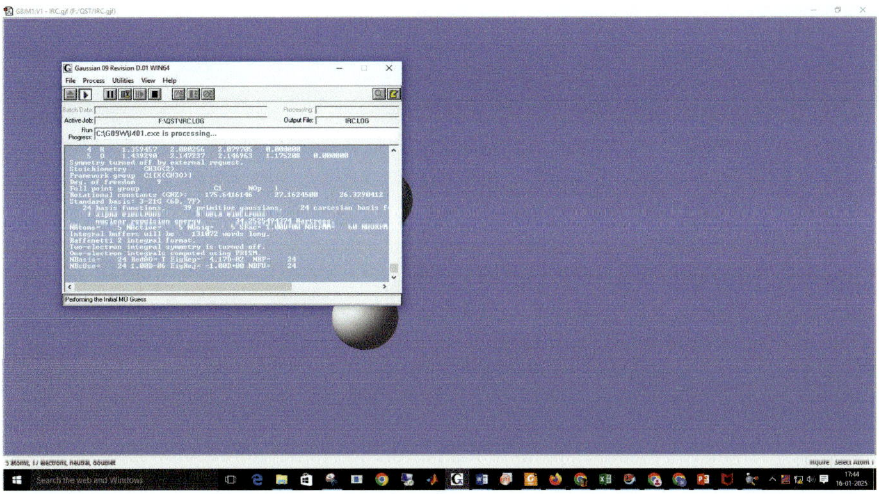

FIGURE 5.20 GAUSSIAN interface showing that the IRC optimization was successfully completed.

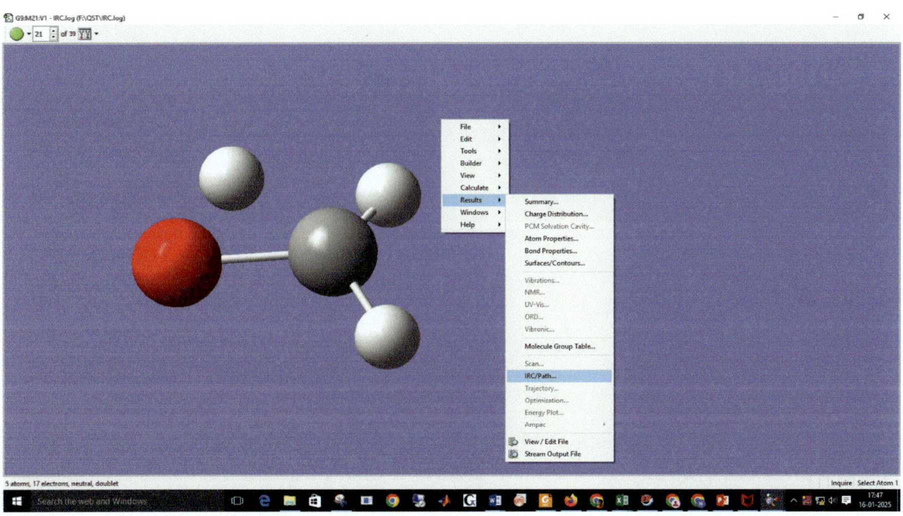

FIGURE 5.21 GAUSSIAN result section displaying the option to access IRC results.

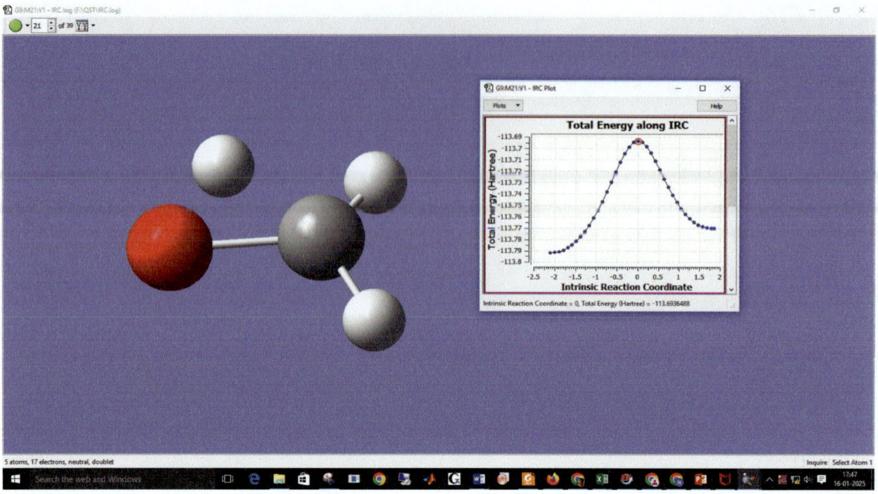

FIGURE 5.22 Graphical representation of the calculated IRC path connecting reactants and products via the transition state.

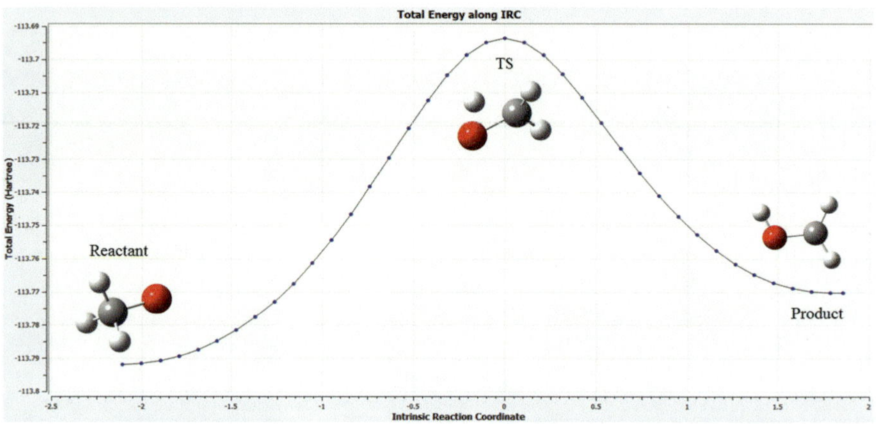

FIGURE 5.23 Final visualization showing the complete reaction pathway between reactants and products, confirming the smooth transition via the transition state.

REFERENCES

1. Roca, M. and I.H. Williams, Transition-state vibrational analysis and isotope effects for COMT-catalyzed methyl transfer. *Journal of the American Chemical Society*, 2020. 142(36): p. 15548–15559.
2. Dhongde, N.R., N.K. Das, and J, Hazarika, et al., Azoles as corrosion inhibitors in alkaline medium for ruthenium chemical mechanical planarization applications: Electrochemical and theoretical analysis. *Journal of Molecular Structure*, 2025. 1320: p. 139651.

3. Gandour, R., *Transition States of Biochemical Processes*. Springer Science & Business Media, 2013, US.
4. Dhongde, N.R., N.K. Das, and T. Banerjee, Synthesis of carbon quantum dots from rice husk for anti-corrosive coating applications: Experimental and theoretical investigations. *Industrial Crops and Products*, 2024. 212: p. 118329.
5. Frisch, M., Gaussian 09, Revision D. 01, Gaussian Inc., 2009, Wallingford, USA.
6. GaussView, Version 3.09, R. Dennington II, and T. Keith, et al., *Gaussian software*, Semichem Inc.: Shawnee Mission, KS, 2003.
7. Moreno, C., On the spaces of positive and negative frequency solutions of the Klein-Gordon equation in curved space-times. *Reports on Mathematical Physics*, 1980. 17(3): p. 333–358.

6 Introduction to Molecular Dynamics Simulation

6.1 MOLECULAR DYNAMICS SIMULATION

6.1.1 Introduction

Molecular dynamics (MD) simulation is an advanced computer method that has transformed the study of atomic and molecular behavior. MD simulations provide distinct atomic-level insights into the dynamic mechanisms that dictate the behavior of materials and biomolecules by computationally solving Newton's equations of motion for a system of interacting particles. This technique allows researchers to monitor the temporal progression of molecular systems, recording events such as protein folding, chemical interactions, diffusion, and phase transitions. The usefulness of MD is in its capacity to simulate the intricate interactions of forces inside a system, enabling the prediction of properties and behaviors that are often difficult to determine experimentally. MD simulations are essential for investigating molecular interactions, innovating materials, and formulating new pharmaceuticals in chemistry, materials science, biology, and pharmacology. The insights derived from these simulations augment our theoretical comprehension and propel substantial advancements in several scientific and technical fields.

6.1.1.1 Molecular Dynamics Methodology

MD simulation consists of the numerical, step-by-step, solution of the classical equations of motion, which for a simple atomic system may be written as

$$f_i = m_i \frac{d^2 r_i}{dt^2} \tag{6.1}$$

where f_i is the force, m_i is the mass, and r_i is the acceleration of atom i. The potential energy $U(\vec{r}_N)$, where $\vec{r}_N = (r_1; r_2; \ldots r_N)$ denotes the full set of $3N$ atomic coordinates, is the source of the forces f_i acting on the atoms. An immediate way to calculate the force acting on atom i is to take the derivative of the potential energy $U(\vec{r}_N)$ with respect to the coordinates:

$$f_i = -\frac{\partial U(r_N)}{\partial r_i} \tag{6.2}$$

The $U(r)$ term represents an externally applied potential also called as *force field* [1], which is used for full periodic simulations of bulk systems.

DOI: 10.1201/9781032694733-6

6.1.1.2 Force Field

The accuracy of a classical molecular simulation in predicting thermophysical parameters is contingent upon the quality of the force field employed to describe fluid interactions. The force field comprises the subsequent components for the computation of energy and force: A compilation of atomic types: (a) a list of atom types, (b) a compilation of atoms, (c) atom-typing regulations, (d) functional representations for the constituents of the energy expression, and (e) parameters for the functional components.

In our work on ionic liquids, we have used the standard molecular mechanics force field [2] with functional form as given below:

$$U_{\text{total}} = \sum_{\text{bonds}} k_b (r - r_0)^2 + \sum_{\text{angles}} k_\theta (\theta - \theta_0)^2 + \sum_{\text{dihedrals}} k_\chi \left[1 + \cos(n\phi - \delta)\right]$$
$$+ \sum_{\text{impropers}} k_\psi (\psi - \psi_0)^2 + \sum_{i=1}^{N-1} \sum_{j>1}^{N} \left\{ \varepsilon_{ij} \left[\left(\frac{\sigma_{ij}}{r_{ij}} \right)^{12} - 2\left(\frac{\sigma_{ij}}{r_{ij}} \right)^{6} \right] + \frac{q_i q_j}{r_{ij}} \right\} \tag{6.3}$$

The force field describing the molecule employs a combination of bonded interaction terms (first four terms in Equation 6.3), such as bond distance r_0, bond angles θ_0, dihedrals δ, and impropers ψ_0 (to maintain the planarity of the ring), which describe a part of the total potential U_{total} due to interactions between bonded atoms.

Together with the respective force constants k_b, k_θ, k_χ, and k_ψ, they are together called *force field parameters* (Figure 6.1). The nonbonded terms include the van der Waals (5th term) better known as *Lennard Jones potential* and electrostatic interactions (6th term) between atoms. In the force field equation (Equation 6.3), the number 2 in *LJ potential* term is used as per NAno scale Molecular Dynamics (NAMD) [3a,b] convention. The *LJ* parameters for unlike atoms are estimated using the Lorentz–Berthelot rule, i.e.,

$$\sigma_{ij} = \frac{\sigma_i + \sigma_j}{2} \quad \varepsilon_{ij} = \sqrt{\varepsilon_i \varepsilon_j} \tag{6.4}$$

Extensive literature by Maitland et al. [4], Gray and Gubbins [5], and Stone [6] indicates how these potentials are determined experimentally or modeled theoretically. Calculating the forces exerted on each atom is necessary for function evaluations in

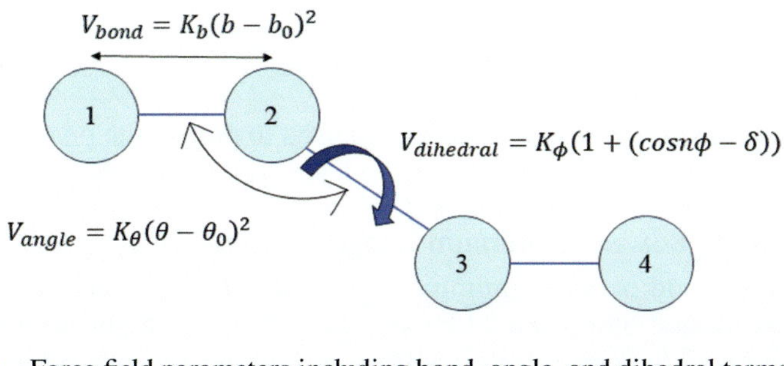

FIGURE 6.1 Force field parameters including bond, angle, and dihedral terms.

Note: 1, 2, 3, 4: Atom indices; $_0$ refers to their equilibrium position

MD. Force calculations in MD simulations need a disproportionately large amount of processing power (up to 90%) and must be executed just once every time step. Because intramolecular bonds can have extremely high vibration frequencies, they are not always necessary to include them in the potential energy function when calculating the force field. Rather, the bonds are considered to be limited to a constant length. For larger molecules (e.g., proteins or polymers) in condensed phases, families of force fields like AMBER [7], CHARMM [8], and OPLS are better suited; these fields often use quantum chemical calculations in conjunction with thermophysical and phase coexistence data to determine their parameters, and their functional forms are simpler.

6.1.1.3 Integrating Algorithms

For solving the equations of motion, the usual approach is to integrate a set of first-order differential equations (Equation 6.5). This is required to advance the system variables through a discrete step in time δt by finite differences approximation:

$$\ddot{r}_i = \frac{\vec{p_i}}{m_i}, \quad \dot{p}_i = \mathbf{F}_i \tag{6.5}$$

For these function evaluations in MD, the forces acting on each atom must be computed. A further characteristic of a good integrator is the ability to undo its effects over time. This implies that the system would theoretically retrace its previous trajectory if the velocities were to ever be reversed. The Verlet technique, developed by Verlet [9], is a popular technique that uses a differential time step to simply expand the atomic coordinates forward and backward in time:

$$\begin{aligned} \mathbf{r}(t+\delta t) &= \mathbf{r}(t) + \tfrac{1}{m}\mathbf{p}(t)\delta t + \tfrac{1}{2m}\mathbf{F}(t)\delta t^2 + \ldots \\ \mathbf{r}(t-\delta t) &= \mathbf{r}(t) - \tfrac{1}{m}\mathbf{p}(t)\delta t + \tfrac{1}{2m}\mathbf{F}(t)\delta t^2 - \ldots \end{aligned} \tag{6.6}$$

Addition of both equations leads to:

$$\mathbf{r}(t+\delta t) + \mathbf{r}(t-\delta t) = 2\mathbf{r}(t) + \tfrac{1}{m}\mathbf{F}(t)\delta t^2 + \ldots \tag{6.7}$$

which upon rearrangement gives us a path for the position in the next time step:

$$\mathbf{r}(t+\delta t) = 2\mathbf{r}(t) - \mathbf{r}(t-\delta t) + \tfrac{1}{m}\mathbf{F}(t)\delta t^2 + \ldots \tag{6.8}$$

The position at the previous step is saved and used to project the position at the next step. It should be noted that the positions are updated without taking the velocities.

6.1.1.4 Periodic Box and Minimum Image Convention

Surface effects would be more prominent in a virtual system because the ratio of surface atoms to total atoms would be significantly larger than in reality, and the number of atoms in the system (N) would be insignificant compared to the number of atoms in a macroscopic piece of matter (of the order of 10^{23}). This is accomplished

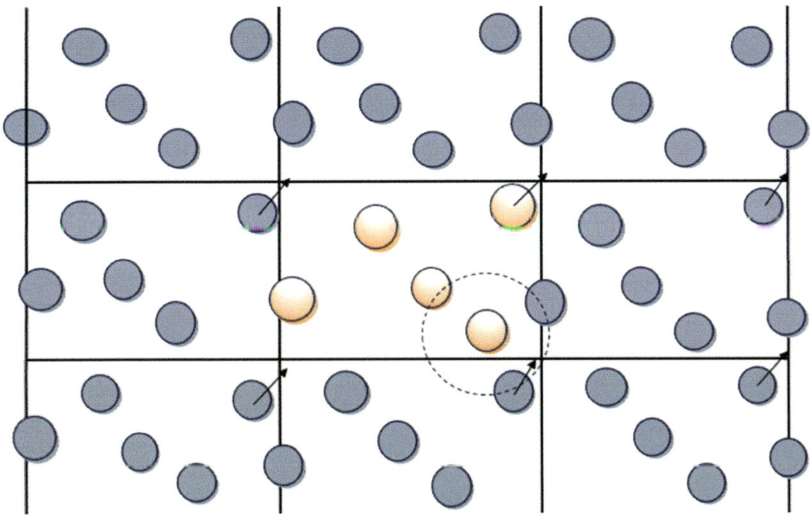

FIGURE 6.2 Periodic box condition (the dotted line circle indicates the cutoff radius (r_c)) (the yellow-colored cell indicates the central cell which is translated in x,y, and z direction).

Note: The colored orange spheres belong to the central box. The images in blue (with arrows) from adjoining box replaces the atoms of the central box. This is replicated in all the boxes.

via the periodic boundary condition (PBC), which entails enclosing particles in a box and then rigidly translating the box in all three Cartesian directions until it fills the space to infinity.

Every atom interacts with the one closest to it in the periodic array, based on the assumption that the potential range is small. If one atom departs from the fundamental simulation box during the course of the simulation, the focus might be shifted to the incoming image. When the distance between two atoms is more than or equal to a certain cutoff distance R_c, we say that the particles are no longer interacting with one another (Figure 6.2). By doing so, we may save time on costly force calculations.

In a system of N atoms, the half-life of each pair is directly related to the number of unique pairings, or ½ $N(N-1)$. To circumvent this, a "skin" is added to the potential cutoff sphere (R_c) around an atom in a pair listing, resulting in a bigger sphere (R_{list}) than before. The initial step is to compile a list of all the atoms that have pairs of neighbours with pair separations inside R_{list}. During the subsequent MD time steps, the force routine will only verify pairs that are present in the list. The list is periodically rebuilt; this must be done before any unlisted pairs cross the safety zone and approach interaction range (Figure 6.3). Every atom pair should not move more than the *pairlistdist* – *cutoff* in a single cycle, hence the pairlist distance value should be selected accordingly. Efficiency and conservation of energy are guaranteed by this.

6.1.1.5 Long Range Forces

A long-range force is defined typically for the charge interaction between ions ($\sim r^{-1}$) and the dipole-dipole (r^{-3}) interaction between molecules. These forces are important considering their range to be greater than half the length of the box. For handling

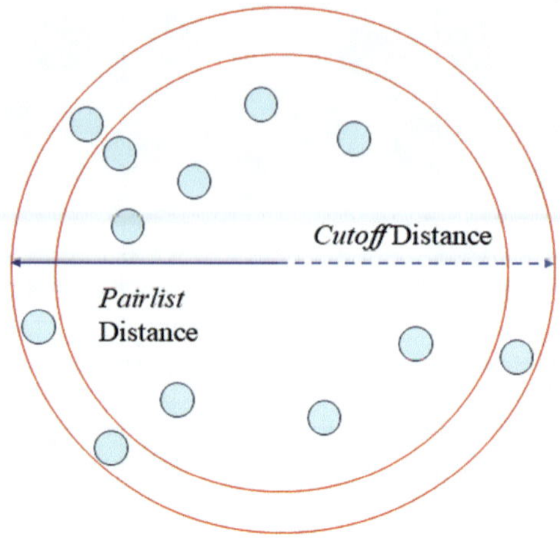

FIGURE 6.3 Cutoff and the pair list distance (the oval shape indicates atoms).

Note: Atom pairs are allowed to move in and out of the cutoff distance without causing energy conservation to be disturbed.

such force Ewald Sum method is employed, which efficiently sums the interaction between an ion and all its periodic images. The potential can be written as:

$$U_{zz} = \frac{1}{2}\sum_{n}{}' \left(\sum_{i=1}^{N} \sum_{j=1}^{N} z_i z_j \left| r_{ij} + n \right| \right) \tag{6.9}$$

Here z_i, z_j are the charges. The sum over n is the sum over all simple lattice points, i.e., $n = (n_x L, n_y L, n_z L)$, where n_x, n_y, and n_z are integers. Omitting $i = j$ for $n = 0$, the sum becomes conditionally convergent. The unit cells are added in sequence, i.e., *n = 0; n = (0, 0, 0); the second term n = L, i.e.,* we have six boxes given by $n = (\pm L, 0, 0)$, $(0, \pm L, 0)$, $(0, 0, \pm L)$. In this way, an infinite set of roughly spherical layers is built. Here, the nature of the medium surrounding the sphere, i.e., the relative permittivity ε_s should be mentioned. Thus, we have

$$U_{zz}\left|(\varepsilon_s = \infty) = U_{zz}\right|(\varepsilon_s = 1) - \frac{2\pi}{3L^3}\sum_{i}(z_i r_i)^2 \tag{6.10}$$

where $(\varepsilon_s = 1)$ is the permittivity for a sphere surrounded by vacuum. The mathematical details can be found in De Leeuw et al. [10] and Heyes [11].

6.1.1.6 Nonbonded Interaction

6.1.1.6.1 Nonbonded van der Waals Interactions

The cutoff distance (R_c) always truncates the nonbonded van der Waals interactions. Among the available options, the switching parameter has the most impact on van der Waals interactions. To smoothly truncate the van der Waals potential energy at the cutoff distance, we can enable this option and a smooth switching function will be employed. As seen in Figure 6.4, the van der Waals potential can be visualized using this switching function. At the cutoff distance, the potential is smoothly

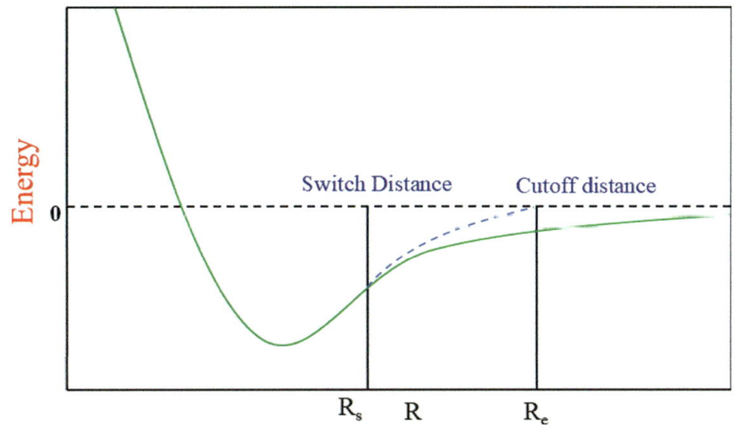

FIGURE 6.4 Effect of non-bonded interaction without/with switching including switch distance and cut-off distance.

Note: *Switchdist* in NAMD specifies the distance at which the switching function is activated in order to converge the van der Waals potential to zero by the cutoff distance.

lowered to zero when the switching function is enabled. A truncation of the potential occurs in the absence of the switching function. Therefore, switchdist (R_s) must always have a lower value than cutoff.

6.1.1.6.2 Multiple Time Stepping (MTS)

A MTS integration approach is employed to further decrease the computational cost of the complete electrostatics [1]. A local component that changes very fast and a long-range component that changes much more slowly make up the overall force exerted on each atom in this approach. Equation 6.11 provides a splitting function that is used to define the local force component:

$$
\begin{aligned}
&SW(r_{ij}) = 0 && |r_{ij}| > R_c \\
&SW(r_{ij}) = 1 && |r_{ij}| \le R_s \\
&SW(r_{ij}) = \frac{\left(R_c^2 - |r_{ij}|^2\right)^2 \left(R_c^2 + 2|r_{ij}|^2 - 3R_s^2\right)}{\left(R_c^2 - R_s^2\right)^3} && R_c > |r_{ij}| > R_s
\end{aligned}
\tag{6.11}
$$

Here, the switch distance is R_s and the cutoff distance is R_c. All electrostatic interactions, including those between bounded and van der Waals pairs, and those between pairs with a separation smaller than the local interaction distance (as specified by the splitting function) make up the local force component. Electrostatic interactions beyond the local interaction distance are the sole components of the long-range component. The long-range forces are not assessed at each timestep due to their gradual variation. The NAMD program, which stands for "NAno scale Molecular Dynamics" [3a,b], instead evaluates them every k timesteps. At each of

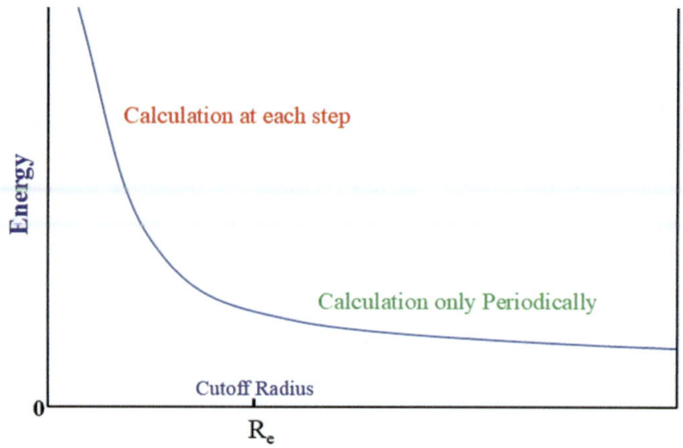

FIGURE 6.5 Electrostatic potential with full electrostatics.

Note: Forces are divided into quickly varying local component (bonded and vdW/electrostatic separated by less than local interaction distance) and a slow long range component (electrostatic interactions outside of the local interaction distance. This is specified by *fullElectFrequency*.

the k timesteps, the system is subjected to the long-range force using the r-RESPA integrator, as described by Tuckerman et al. [12]. The simulations are significantly sped up because less forces need to be calculated at each time step. Cutting down on the number of pairs of forces that need to be evaluated results in a considerable decrease in CPU time, which is a major factor in system simulation time due to the computation of the forces.

The local interaction distance still truncates the van der Waals forces in this design. Therefore, the local interaction distance has a minimum value equal to the van der Waals cutoff distance. Because of this cutoff distance, we know exactly which electrostatic pairs are calculated at each timestep. Calculations of interactions will only be performed regularly beyond this distance, as shown in Figure 6.5.

6.1.1.7 Temperature Control

Keeping the temperature constant is essential for NVT and NPT ensembles. For constant temperature dynamics, one can use Nosé dynamics [13]. Equation 6.12 describes the method for controlling temperature, which involves adding a new "thermal reservoir" variable to the dynamical equations. To obtain within 3% of the target temperature, the Verlet method requires a timestep of 0.5 fs. As the timestep decreases, the goal temperature is approached more closely:

$$\dot{p}_i = f_i - \xi \mathrm{p_i}$$

$$\dot{\xi} = \frac{\sum_{i\alpha} \frac{p_i^2}{m_i} - NkT}{W} = v_T^2 \left[\frac{\sum_{i\alpha} \frac{p_i^2}{m_i}}{NkT} - 1 \right] = v_T^2 \left[\frac{\tau}{T} - 1 \right] \tag{6.12}$$

Here, the friction coefficient namely ξ is allowed to vary in time. W is the thermal inertia parameter, which is replaced by v_T. v_T is the decay time for thermal fluctuations. Further, N is the number of degrees of freedom and τ stands for instantaneous "mechanical temperature". If $\frac{\tau}{T} > 1$, then the system temperature is hot, thus the friction coefficient ξ will increase and vice versa.

6.1.1.7.1 Langevin Dynamics

It is highly implausible that a molecular system will exist in a vacuum in the real world. Friction is generated by the interaction of molecules, and the system may be perturbed by the occasional high-velocity collision. Langevin dynamics endeavours to expand the scope of mathematical dynamics to accommodate these effects. This simulates the viscous effect, which is comparable to the procedure described above. NAMD is capable of executing Langevin dynamics or temperature control, which involves the introduction of random forces and attenuation to the system. The desired temperature (*langevinTemp*) and the damping coefficient (*langevinDamping*) ξ ps^{-1}, in conjunction with the keyword *langevin*, are necessary for temperature control.

6.1.1.8 Pressure Control

PBCs are essential for constant pressure simulation. Pressure is regulated by dynamically adjusting the dimensions of the cell and rescaling all atomic coordinates throughout the simulation, as outlined by Feller et al. [14]. In NAMD [3a,b], the constant pressure method described by Martyna et al. [15] is implemented. This method incorporates an additional degree of freedom, denoted as s, which corresponds to the volume of the cubic simulation cell. This degree of freedom adjusts to ensure that the internal and applied pressures are equalized. This degree of freedom functions as a piston and is assigned a "mass" W, which has units of mass per length to the fourth power. This degree of freedom simulates the action of a piston in a real system. This pertains to kinetic energy in relation to velocity V:

$$K_v = \frac{1}{2} W\dot{V}^2 \tag{6.13}$$

Thus, we have:

$$\frac{dP}{dt} = \frac{(P_d - P)}{t_p} \tag{6.14}$$

Here, P_d is the desired pressure and t_p is the time constant for pressure fluctuations and $\beta = 1/kT$. At each step, the volume of the cell is scaled by a factor of χ and the molecular centre of mass by a factor $\chi^{0.33}$. Thus, we have:

$$\chi = 1 - \beta \frac{\delta t}{t_p}(P_d - P) \tag{6.15}$$

This method, when combined with a method of temperature control, creates trajectories in the NPT ensemble., For the NPT ensemble Hoover [16] gave the following equations for pressure control:

$$\dot{s} = \frac{p}{mV^{1/3}} \quad \text{(a)} \quad \dot{p} = f - (\chi + \xi)p \quad \text{(b)}$$

$$\chi = \frac{\dot{V}}{3V} \quad \text{(c)} \quad \dot{\chi} = \frac{(P - P_d)V}{t_p^2 kT} \quad \text{(d)} \tag{6.16}$$

Here, ξ is given by Equation 6.16b, k is the Boltzmann constant, and f is the frictional force. Here we have to specify the desired pressure P_d, time constant t_p, decay time v_T and instantaneous temperature τ of the piston. In addition, the damping coefficients ξ and desired temperature of the molecule T are also specified for maintaining the constant temperature. For bulk solvent/fluid, the instantaneous pressure of a simulation cell will have mean square fluctuations of $kT/(V\beta)$ where β is the isothermal compressibility, having root mean square (RMS) of 100 bar for 10,000 atom bimolecular system [17].

6.1.1.9 Estimation of Pure Component Properties

6.1.1.9.1 Specific Heat Capacity

The heat capacity is an example of a "second-derivative" property, in that it can be expressed as a second-derivative of a thermodynamic potential [1]:

$$C_v = \left(\frac{\partial E}{\partial T}\right)_{V,N} = -k\beta^2\left(\frac{\partial^2(\beta A)}{\partial \beta^2}\right)_{V,N} \tag{6.17}$$

The formula for evaluating it by molecular simulation follows in a simple way from the expression for the average energy. Taking $\beta = 1/kT$, **a** simple expression for the heat capacity results:

$$C_v = k\beta^2\left[\langle E^2\rangle - \langle E\rangle^2\right] \tag{6.18}$$

The heat capacity is given in terms of the variance of the distribution of energies in the canonical ensemble, i.e., NVT ensemble. A broad distribution of energies corresponds to a large heat capacity.

6.1.1.9.2 Volume Expansivity

The volume expansivity quantifies the extent to which the volume of a fluid changes with temperature at constant pressure, and is defined as:

$$\alpha_p = \frac{1}{V}\left(\frac{\partial V}{\partial T}\right)_P \tag{6.19}$$

This can be found out by running a number of simulations in a NPT ensemble at different temperatures.

6.1.1.9.3 Isothermal Compressibility

The second-derivative properties are expressed as variances or covariances of the corresponding first-derivative properties. Thus, we have the compressibility given as the variance of the volume in the NPT ensemble [1]:

$$\begin{aligned}\kappa_T &\equiv -\frac{1}{V}\left(\frac{\partial V}{\partial P}\right)_{T,N} \\ &= \beta\frac{1}{V}\left[\left\langle V^2\right\rangle - \left\langle V\right\rangle^2\right]\end{aligned} \tag{6.20}$$

6.1.1.9.4 Cohesive Energy Density and Solubility Parameter

The cohesive energy density of a liquid is defined by Prausnitz et al. [18]:

$$\mathrm{CED} = \frac{\Delta IE^{\mathrm{vap}}}{V^L} \quad \text{(a)} \quad \delta = \sqrt{\frac{\Delta IE^{\mathrm{vap}}}{V^L}} \quad \text{(b)} \tag{6.21}$$

where ΔIE^{vap} represents the change in internal energy of the liquid during isothermal vaporization into the ideal gas state, while denotes the molar volume of the liquid. To determine the cohesive energy density (Equation 6.21a), the average internal energy and molar volume of the liquid were derived from the liquid simulation. Additionally, the average internal energy of the ideal gas was acquired by simulating a single ionic liquid ion pair at the same temperature as the liquid, while maintaining zero pressure (i.e., no PBCs). The solubility parameter, which predicts the extent of interaction between molecules, is expressed by Equation 6.21b.

6.1.1.9.5 Root Mean Square Deviation

The coefficient of self-diffusion for a fluid or solid can be calculated using the Einstein relation [19]:

$$\mathrm{RMSD}(t) = \sqrt{\frac{1}{N}\sum_{i=1}^{N}\left[r_i(t) - r_{\mathrm{ref},i}\right]^2} \tag{6.22}$$

$$\mathrm{MSD}(t) = \left\langle \left[r_i(t) - r_i(0)\right]^2 \right\rangle$$

$$\mathrm{RMSD}(t) = \sqrt{\mathrm{MSD}(t)}$$

where the quantity in braces is the ensemble-averaged mean square displacement (MSD) of the molecules and r_i is the vector coordinate of the center of mass of the ion/molecule.

6.1.1.9.6 Coefficient of Self-Diffusivity

The coefficient of self-diffusion for a fluid or solid can be calculated using the Einstein relation [19]:

$$D_{\text{self}} = \frac{1}{6}\lim_{t\to\infty}\frac{d}{dt}\left\langle \left| r_i(t) - r_i(0) \right|^2 \right\rangle \tag{6.23}$$

where the quantity in braces is the ensemble-averaged mean square displacement (MSD) of the molecules and r_i is the vector coordinate of the center of mass of the ion/molecule.

6.1.1.9.7 Ionic Conductivity

The ionic conductivity of the ionic liquid can be found using the Nernst–Einstein relation [20,21]:

$$\sigma_c = \frac{Ne^2 D_{\text{self}}}{VkT} \tag{6.24}$$

where e is the electron charge, V *is* the volume of the simulation cell, and N *is* the number of ions/molecules in the simulation cell.

6.1.1.10 Radial Distribution Function

In MD, molecules exhibit continuous motion, characterized by rotation and erratic movement. Consequently, the concept of structure is relevant primarily in an average context. Multiple methodologies exist for quantifying this average structure. The radial distribution function (rdf) is a significant method and is among the most critical techniques in this context. The radial distribution function (rdf) quantifies the expected number of atoms located at a distance r from a specific atom, specifically within the range of r to r+dr (refer to Figure 6.6). The radial distribution function (rdf) is a fundamental quantity in statistical mechanics, as it defines the manner in which atomic correlations diminish with increasing separation distance. The rdf is defined as follows:

$$g(r) = \frac{\rho(r)d\mathbf{r}}{\rho^{id}d\mathbf{r}} \tag{6.25}$$

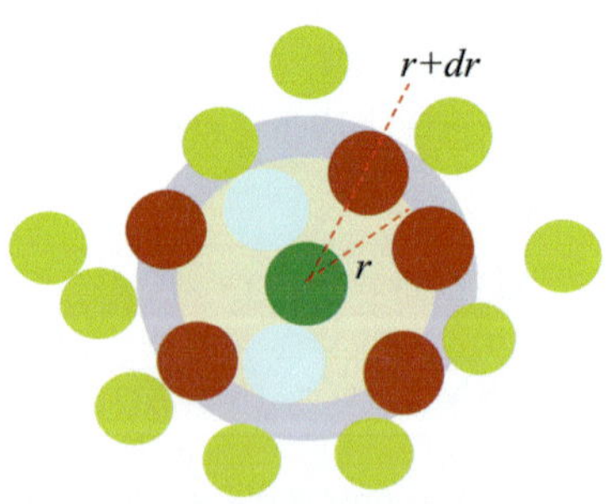

FIGURE 6.6 Schematic representation of radial distribution function (RDF).

Note: The central atom with green sphere is being compared to the spheres within a vicinity from r to $r+dr$. The blue region resembles the area/volume with the differential radius (dr). The brown spheres lie in the differential volume.

The numerator represents the count of atoms located within a volume element dr at a distance r from a specified atom. Conversely, the denominator denotes the equivalent quantity for an ideal gas, which is characterized by the absence of atomic correlations. The ideal gas term exhibits independence from the variable r and is defined as the number density. The understanding of rdf is essential for the calculation of thermodynamic properties, specifically energy and pressure. The rdf can be quantified through experimental methods, specifically employing neutron-scattering techniques.

6.1.1.11 Molecular Dynamics Packages

Popular packages for conducting MD simulations include CHARMM (Chemistry at HARvard Macromolecular Mechanics) [8], assisted model building and energy refinement (AMBER) [7], GRoningenMAchine for Chemical Simulations (GROMACS) [22], and NAMD [3a,b]. AMBER refers to a comprehensive suite of software designed for conducting and analyzing MD simulations, with a focus on proteins, nucleic acids, and carbohydrates. This represents a collection of force fields utilized for the MD simulation of biomolecules.

CHARMM is a software application designed for the simulation of macromolecular dynamics and mechanics. The software executes standard MD simulations across various ensembles, such as NVE, NVT, and NPT. It employs advanced algorithms for time stepping, long-range force calculations, and the management of periodic images. This tool is applicable for energy minimization, normal mode analysis, and crystal optimizations. The potential energy functions implemented in CHARMM have undergone thorough parameterization for the simulation of proteins, nucleic acids, and lipids. The CHARMM software provides fully developed methods for calculating chemical and conformational free energy using free energy techniques.

GROMACS is specifically engineered for biochemical molecules such as proteins and lipids, which exhibit numerous complex bonded interactions. However, due to its high efficiency in calculating nonbonded interactions—typically the dominant factor in simulations—it is also applicable in research involving nonbiological systems, including polymers. The system is designed for ease of use, featuring topologies, parameter files, and error messages presented in a clear text format. Consistency checking is extensively implemented; however, there is no scripting language available. All programs operate through a straightforward interface that utilizes command line options for specifying input and output files.

NAMD is a parallel MD software developed for the efficient simulation of extensive biomolecular systems. The system is capable of scaling to hundreds of processors on high-end parallel platforms and to tens of processors on commodity clusters utilizing gigabit Ethernet. The system is designed for ease of use, featuring topologies, parameter files, and error messages presented in a clear text format. NAMD utilizes the widely recognized molecular graphics software visual molecular dynamics (VMD) for the purposes of simulation setup and trajectory analysis. It is also compatible with the AMBER, CHARMM, and XPLOR [23] force fields. The current version of NAMD is 3.0.1, which enhances compatibility with the OPLS force field as described by Jorgensen et al. [24].

The implementation includes the particle mesh Ewald (PME) algorithm, which considers the complete electrostatic interactions. This algorithm decreases the computational complexity associated with the evaluation of electrostatic forces from O (N^2) to O (N log N). To further reduce the evaluation of long-range electrostatic forces, a multiple time step scheme is utilized. At each time step, the local interactions, including bonded, van der Waals, and electrostatic interactions within a specified distance, are computed. The longer-range interactions, specifically the electrostatic interactions that extend beyond the defined distance, are computed at periodic intervals. This minimizes the computational expense associated with calculating electrostatic forces across multiple timesteps. The file formats for input and output utilized by NAMD are the same as those employed by CHARMM and X-PLOR. Input formats consist of coordinate files formatted in Protein Data Bank (PDB) format [25], structure files formatted in X-PLOR PSF (protein structure file) format, and energy parameter files available in either CHARMM or X-PLOR formats. Output formats consist of PDB coordinate files and binary DCD trajectory files. The identified similarities ensure compatibility between MD trajectories generated by NAMD and those utilized by CHARMM or X-PLOR, thereby enabling users to leverage the extensive analysis algorithms available in these software packages.

6.1.2 Generating Input Files in Gaussian

For carrying out MD simulation, packages such as CHARMM (**C**hemistry at **HAR**vard **M**acromolecular Mechanics) [8], AMBER (**A**ssisted **M**odel **B**uilding and **E**nergy Refinement) [7], GROMACS (**GRO**ningen**MA**chine for **C**hemical Simulations) [22], and NAMD [3a,b] are currently being used.

6.1.3 Force Field Parameters from AMBER (ANTECHAMBER)

The ANTECHAMBER module within the AMBER suite is intended to generate force field parameters for the organic compounds. ANTECHAMBER optimizes parameter assignment by automating the calculation of atomic charges, bond lengths, angles, and torsions. The procedure often entails:

6.1.3.1 Input Preparation

Employing software such as Gaussian or other quantum chemistry programs to optimize the structure and compute the electrostatic potential (ESP).

For geometry optimization, following commands can be used in the route section of the Gaussian calculation setup: **# opt freq b3lyp/6–311++g(d,p) geom = connectivity pop = chelpg**.

In this calculation, the keywords are

opt: optimization

freq: frequency calculation

b3lyp/6–311++g(d,p):triple-zeta split-valence basis set with additional diffuse functions (++) and polarization functions (*d*, *p*), enhancing the accuracy of quantum chemical calculations for atoms and molecules.

pop=chelpg: CHELPG (Charges from ESPs using a Grid-based method) and population analysis (e.g., Mulliken) are computational chemistry methodologies used to determine atomic charges via the fitting of electron density or ESP distributions.

Preparation of RESP output file:

%nprocshared=2

%mem=150MW

%chk=E:\BOOK-REVIEW\BOOK2-QC-MD\MD\Acetone-methanol\ACN.chk (directory location of the .chkfile which will be different for the individual users)

 # opt freq b3lyp/6-311+g(d) geom=connectivity· route section for geometry optimization

-Link1--

%nprocshared=2

%mem=150MW

%chk=E:\optimization\Ethylene glycol\ethy_gcl.chk

 # HF/6-31G* SCF=Tight Geom=AllCheck Guess=Read Pop=MK IOp (6/33=2,6/41=10,6/42=17)

RESP calculation

0 1

Here is an elucidation of each component of the command, emphasizing the internal alternatives.

HF/6–31G*: This denotes a Hartree-Fock (HF) computation using the 6-31G* basis set, characterized as a split-valence basis set inclusive of polarization functions.

SCF=Tight: This establishes rigorous convergence conditions for the Self-Consistent Field (SCF) process, indicating that computations will persist until a higher degree of precision in energy and density is attained.

Geom=AllCheck: This specifies that the geometry of the molecule should be read from a checkpoint file created by a prior computation.

Guess=Read: This command directs the software to get the initial guess for the wavefunction from the checkpoint file, facilitating expedited convergence.

Pop=MK: This indicates that a Merz–Kollman (MK) population analysis is to be conducted, which determines atomic charges by fitting to the ESP.

IOp(6/33=2): This internal option determines the approach for managing two-electron integrals, with the value 2 generally indicating a specific storage or processing technique (e.g., direct integral storage).

IOp(6/41=10): This internal option alters the SCF convergence parameters. The number 10 often signifies a certain convergence threshold, so making the convergence process either more rigorous or more permissive.

IOp(6/42=17): This internal option influences the SCF density matrix updating methodology. The number 17 often denotes a specific approach or procedure used to modify the density matrix during SCF rounds, influencing both stability and convergence speed.

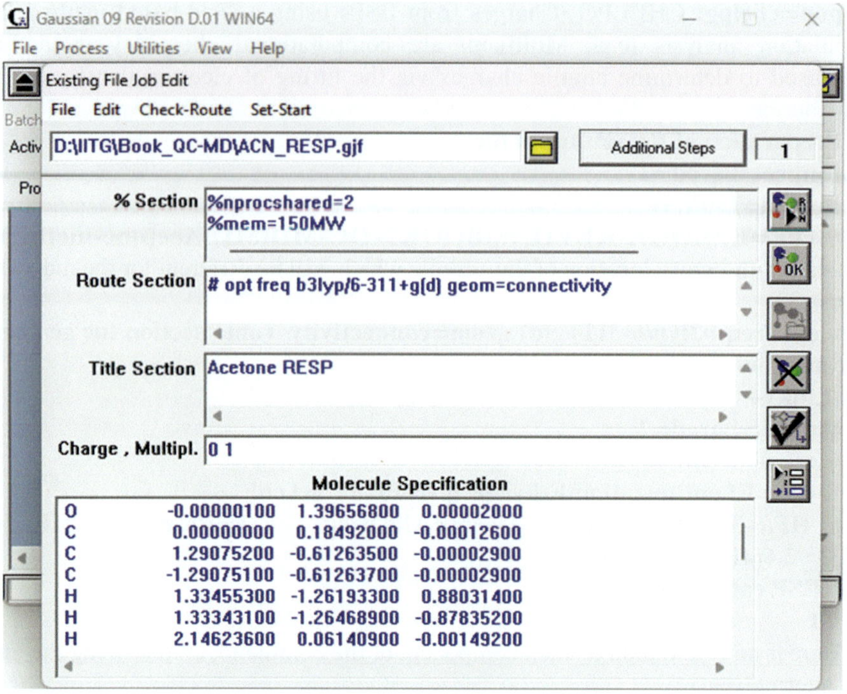

FIGURE 6.7 RESP calculation setup in Gaussian.

Step 1: Save the optimized geometry file (.log) in **(.gjf)** format → Open the **(.gjf)** file in **Gaussian** → type the **route section** for optimization → in **Title Section** write any title **(Acetone RESP)** → it will appear as Figure 6.7.

The following are the steps for computing the restricted electrostatic potential (RESP) calculation from the optimized geometry file.

Step 2: Click on **Additional Steps** → Click on **Add Step→Additional Job Step # 1 of 1** window will appear → type the **route section** for RESP calculation → Click on **OK** → Click on **Run** → it will prompt to save an input file → Save as **ACN_RESP.gjf→** Finally, after completion of the calculation, an output file **(ACN.out)**will be generated. Refer to Figure 6.8.

After the successful generation of RESP output file (ACN.out), the following tasks to be done.

6.1.3.2 Parameter Assignment

Employing ANTECHAMBER to designate atom kinds and provide force field parameters derived from established libraries.

Following is an example of the command steps for generating files such as .mol2, .pdb, and .frcmod for the acetone molecule (Figure 6.9).

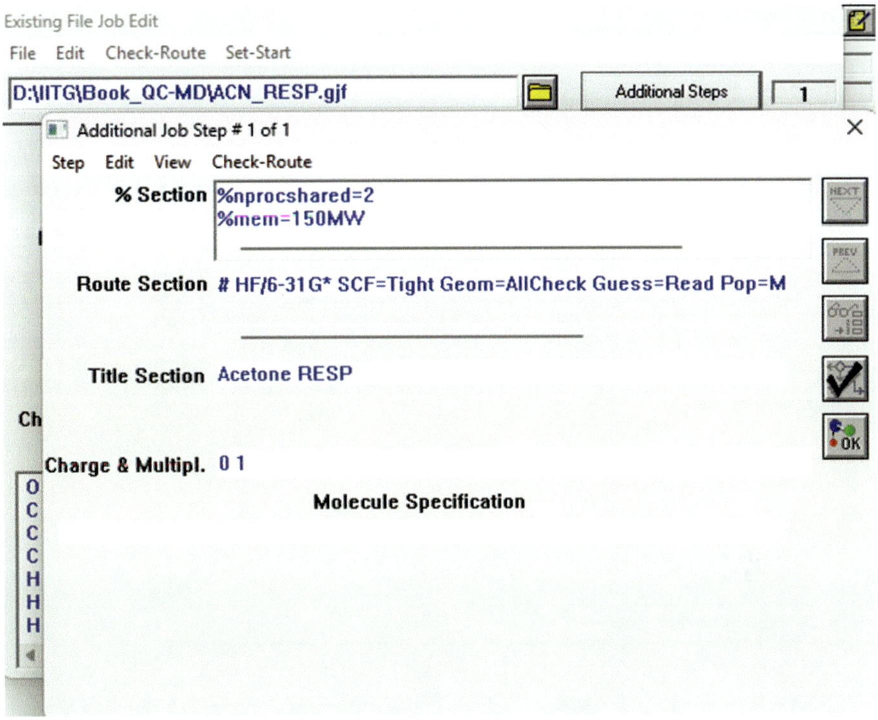

FIGURE 6.8 Additional input step in RESP calculation setup in Gaussian.

```
#!/bin/sh

# Prerequisite: RESP calculation is done in Gaussian 09 and log file is generated

WORKDIR=/home/deb/Desktop/Pesticide
# Give the gaussian file name
#rename the file. Give 3 letter filename you want. This name will be the residue in pdb file.
fname=ACN

cd $WORKDIR

# Antechamber commands.
antechamber -i $fname.out -fi gout -fo mol2 -o $fname.mol2 -c resp -j 4 -rn $fname -pf yes
antechamber -i $fname.out -fi gout -fo prepi -o $fname.prepi -c resp -j 4 -rn $fname -pf yes
antechamber -fi gout -fo ac -i $fname.out -o $fname.ac -c resp -pf yes

# Three important steps. generate atomtype, prepc and modified force field file
atomtype -i $fname.ac -o $fname_gaff.ac -p gaff
prepgen -i $fname_gaff.ac -o $fname.prepc -f car -rn $fname
parmchk -i $fname.prepc -o $fname.frcmod -f prepc

mv NEWPDB.PDB $fname.pdb

rm QOUT qout esout punch *.prepi *.INF *.prepc *.ac
```

FIGURE 6.9 ANTECHAMBER commands in AMBER suite for generation of input files for NAMD calculation [26].

```
@<TRIPOS>MOLECULE
ACN
    10     9     1     0     0
SMALL
resp


@<TRIPOS>ATOM
      1 O1          0.0000     1.3970    -0.0000 o          1 ACN      -0.577955
      2 C1          0.0000     0.1860    -0.0000 c          1 ACN       0.799237
      3 C2          1.2910    -0.6130     0.0000 c3         1 ACN      -0.495934
      4 C3         -1.2910    -0.6130     0.0000 c3         1 ACN      -0.495934
      5 H1          1.3390    -1.2630     0.8790 hc         1 ACN       0.128431
      6 H2          1.3380    -1.2660    -0.8770 hc         1 ACN       0.128431
      7 H3          2.1480     0.0590    -0.0010 hc         1 ACN       0.128431
      8 H4         -2.1480     0.0590    -0.0010 hc         1 ACN       0.128431
      9 H5         -1.3390    -1.2630     0.8790 hc         1 ACN       0.128431
     10 H6         -1.3380    -1.2660    -0.8770 hc         1 ACN       0.128431
@<TRIPOS>BOND
     1     1     2 2
     2     2     3 1
     3     2     4 1
     4     3     5 1
     5     3     6 1
     6     3     7 1
     7     4     8 1
     8     4     9 1
     9     4    10 1
@<TRIPOS>SUBSTRUCTURE
     1 ACN         1 TEMP              0 ****  ****    0 ROOT
```

```
ATOM      1  O1  ACN     1       0.000   1.397  -0.000 -0.577955
ATOM      2  C1  ACN     1       0.000   0.186  -0.000  0.799237
ATOM      3  C3  ACN     1      -1.291  -0.613   0.000 -0.495934
ATOM      4  H4  ACN     1      -2.148   0.059  -0.001  0.128431
ATOM      5  H5  ACN     1      -1.339  -1.263   0.879  0.128431
ATOM      6  H6  ACN     1      -1.338  -1.266  -0.877  0.128431
ATOM      7  C2  ACN     1       1.291  -0.613   0.000 -0.495934
ATOM      8  H1  ACN     1       1.339  -1.263   0.879  0.128431
ATOM      9  H2  ACN     1       1.338  -1.266  -0.877  0.128431
ATOM     10  H3  ACN     1       2.148   0.059  -0.001  0.128431
```

```
remark goes here
MASS

BOND

ANGLE

DIHE

IMPROPER
c3-c3-c -o         10.5          180.0          2.0          General improper torsional angle (2 general atom types)

NONBON
```

FIGURE 6.10 Sample .mol2, .pdb, and .frcmod files for acetone molecule.

The generated mol2, pdb, and frcmod files for the acetone molecule are presented in Figure 6.10a–c.

6.1.3.3 Validation

Confirm that the parameters precisely reflect the molecular characteristics by comparing them with experimental data or advanced computations.

```
tolerance 2.0
filetype pdb
seed 184588
output SystemBox1.pdb

structure ACN.pdb
number 500
add_amber_ter
inside box 0. 0. 0. 50. 50. 50.
end structure

structure MET.pdb
number 500
add_amber_ter
inside box 0. 0. 0. 50. 50. 50.
end structure
```

FIGURE 6.11 Input file for packmol (pack.in): here the system contains 500 molecules of acetone and methanol each.

6.1.4 Generating System Box in PACKMOL

Constructing a System Box in PACKMOL entails constructing a simulation box and arranging molecules in a designated configuration for molecular simulations. This is often used to generate input files for MD or Monte Carlo simulations. PACKMOL [27] enables the specification of box dimensions, molecular types, and quantities, as well as their arrangement inside the box while preventing overlap.

To run this pack.in script, the following command needs to be executed:

cd /path/to/your/directory

packmol< pack.in

The input commands in Figure 6.11 will generate SystemBox1.pdb, which contains a mixed system of acetone and methanol of 500 molecules each. One can change the numbers of the molecules as per the system requirement and can change the coordinates according to that.

6.1.4.1 Procedure for Creating a System Box in PACKMOL

Specify the Dimensions of the Box: Indicate the dimensions (length, width, and height) of the box designated for the placement of the molecules.

Identify Molecules: Enumerate the molecules (e.g., in. pdb or. mol2 format) and specify the number of molecules to be positioned inside the box.

Specify Packing Conditions: Utilize the inner command to position molecules inside the designated box, ensuring no overlap occurs among them.

Execute PACKMOL: Run the PACKMOL application to create the system with molecules organized per the specified input criteria.

The output file, usually in .pdb format, will include the organized system prepared for further simulation (Figure 6.12).

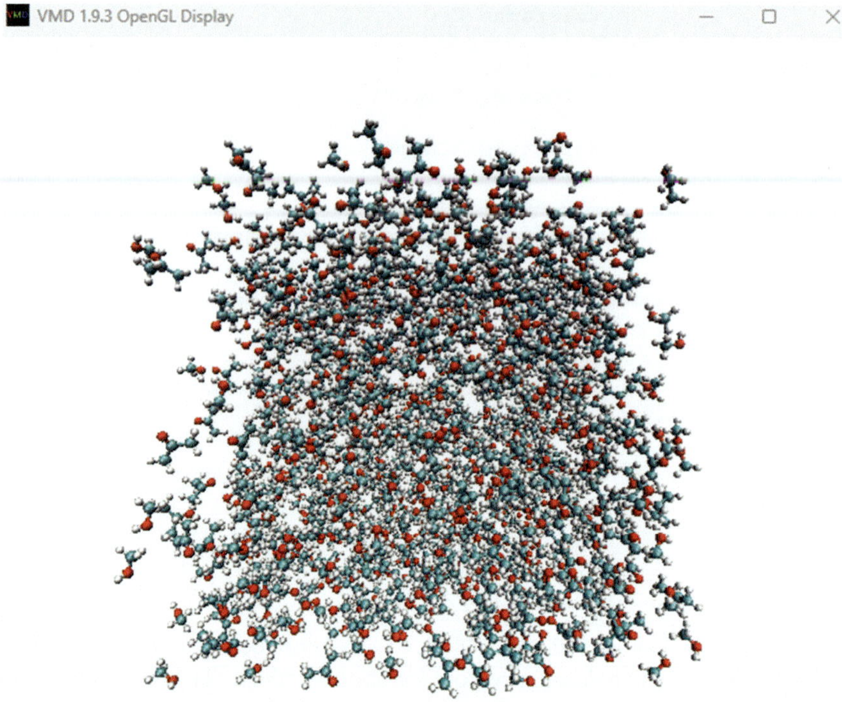

FIGURE 6.12 Systembox after packing in packmol (acetone: 500, methanol: 500).

6.1.5 Running the LEaP Algorithm

This LEaP script is used subsequent to the generation of a system box using PACKMOL [27]. PACKMOL organizes molecules into a specified volume without overlap, producing a PDB file (SystemBox1.pdb) as output. This script utilizes the PDB file, together with molecular structure files (ACN.mol2, MET.mol2) and their respective force field parameter files (ACN.frcmod, MET.frcmod), to construct a comprehensive simulation-ready system in AMBER. The General AMBER Force Field (GAFF) is used in LEaP to provide parameters for molecules.

The command line mentioned in Figure 6.13 can be broken down as follows:

- source leaprc.gaff

  Imports the GAFF parameters into LEaP. This enables LEaP to use GAFF parameters for the compounds one imports.
- ACN = loadmol2 ACN.mol2

  Imports the molecular data from the ACN.mol2 file and assigns it to the variable ACN. The .mol2 file encompasses the molecular structure and atom types of the molecule.
- MET = loadmol2 MET.mol2

  Imports the molecular data from the MET.mol2 file and assigns it to the variable.

```
source leaprc.gaff

ACN = loadmol2 ACN.mol2
MET = loadmol2 MET.mol2

box = loadPdb SystemBox1.pdb

loadamberparams ACN.frcmod
loadamberparams MET.frcmod

setbox box centers

saveAmberParm box DES_DRUG.prmtop DES_DRUG.inpcrd
quit
```

FIGURE 6.13 Input file for leap algorithm (leap.in).

- box = loadPdb SystemBox1.pdb

  Imports the system box coordinates from the SystemBox1.pdb file and assigns them to the variable box. This PDB file generally includes the configuration of molecules inside the simulation box produced by PACKMOL.
- loadamberparamsACN.frcmod

  Imports the force field parameters from the ACN.frcmod file, which includes alterations pertinent to the ACN molecule. This guarantees precise parameterization in accordance with GAFF.
- loadamberparamsMET.frcmod

  Imports the force field parameters from the MET.frcmod file, which includes changes pertinent to the molecule MET. Establishes the box centers.
- setbox box centers

  Establishes the periodic box for the system by centering it around the specified coordinates. This is essential for PBCs in molecular simulations.
- saveAmberParm box DES_DRUG.prmtopDES_DRUG.inpcrd

  Generates the topology (.prmtop) and coordinate (.inpcrd) files for the system box, designated as **DES_DRUG.prmtop**(parameter file) and **DES_DRUG.inpcrd**(input coordinate file), respectively. These files are crucial for doing MD simulations using AMBER.
- quit

  Exits LEaP.

6.2 NAMD SIMULATION SETUP

6.2.1 Introduction

Nanoscale molecular dynamics (NAMD) is a parallel MD software developed for the high-performance modeling of extensive biomolecular systems. The setup procedure involves configuring the molecular system using tools like VMD [28], TRAVIS package [29], or LEaP from the AMBER suite to produce and enhance structure files (e.g., PDB, PSF) and parameter files (e.g., force field parameters). An NAMD configuration file is then generated, delineating critical simulation parameters such as temperature regulation, pressure management, simulation duration, and output preferences. It is essential to ensure precise references to the beginning coordinates, velocity, and structure files. Conducting the NAMD simulation in a suitable computer environment leverages its parallel processing capabilities to efficiently manage and analyze the simulation trajectory. This careful configuration guarantees a strong and dependable MD simulation, yielding significant insights into the behavior of intricate biomolecular systems.

6.2.2 Minimization

Minimization is the procedure for identifying the local minimal energy configuration of a molecular system. This process mitigates steric conflicts and adverse interactions by systematically modifying atomic locations to decrease the system's potential energy. It guarantees a stable initial configuration for future MD simulations, averting unrealistic pressures from compromising the simulation's stability.

```
#############################################################
## JOB DESCRIPTION                                         ##
#############################################################

# Minimization of 500 molecules of Acetone and 500 molecules
of Methanol

#############################################################
## ADJUSTABLE PARAMETERS                                   ##
#############################################################

ambercoor        ../common/DES_DRUG.inpcrd

set temperature    0
set outputnameDES_DRUG_min
set restart     0

# Continuing a job from the restart files
if {$restart} {
set inputnameDES_DRUG_min
Coordinates        $inputname.restart.coor
Velocities         $inputname.restart.vel   ;# remove the
"temperature" entry if you use this!
extendedSystem         $inputname.xsc
}

firsttimestep      0
```

```
##############################################################
## SIMULATION PARAMETERS                                     ##
##############################################################

# Input
amber  on
parmfile      ../common/DES_DRUG.prmtop
if {$restart-1} {
temperature       $temperature
}
```

Explanation:

amber: Indicates that the AMBER force field is being used.

parmfile: Specifies the parameter/topology file in AMBER format.

if {$restart-1}: Another conditional block related to restarts, ensuring temperature is set if not restarting from velocities.

```
# Force-Field Parameters
dielectric 1.0      # Value of the dielectric constant and Any
value greater than 1.0 will be experimental
exclude scaled1-4
nonbondedScaling 1.0
1-4scaling 1.0
cutoff 12.0  #Angstorm
switching on
switchdist 10.0     #Angstorm
pairlistdist 14.0   #Angstorm
```

Explanation:

dielectric: Sets the dielectric constant for the system (1.0 means no dielectric screening).

exclude: Defines which interactions to exclude. Scaled 1–4 means scaling 1–4 interactions according to AMBER rules.

nonbondedScaling: Scaling factor for nonbonded interactions.

1–4scaling: Scaling factor for 1–4 interactions.

cutoff: Nonbonded interaction cutoff distance (12.0 Å).

switching: Enables switching function for nonbonded interactions.

switchdist: Distance at which the switching function begins (10.0 Å).

pairlistdist: Distance within which atom pairs are listed for nonbonded interactions (14.0 Å).

```
# Integrator Parameters
timestep           1.0       ;# 1fs/step
rigidBonds         all       ;# Apply SHAKE algorithm to all
covalent bonds involving hydrogens
rigidTolerance     0.00001   ;# Desired accuracy in
maintaining SHAKEed bond lengths
```

```
rigidIterations      100        ;# Maximum number of SHAKE
iterations
nonbondedFreq        1
vdwGeometricSigma    yes
fullElectFrequency   2
stepspercycle        20
pairlistsperCycle    2
```

Explanation:

timestep: Sets the integration time step to 1.0 femtoseconds.

rigidBonds: Uses the SHAKE algorithm to constrain all covalent bonds involving hydrogens.

rigidTolerance: Accuracy of SHAKE algorithm.

rigidIterations: Maximum iterations for SHAKE algorithm.

nonbondedFreq: Frequency of calculating nonbonded interactions.

fullElectFrequency: Frequency of full electrostatic calculations.

stepspercycle: Number of steps per cycle for calculating interactions.

pairlistsperCycle: Frequency of updating pair lists.

```
# Periodic Boundary Conditions
if {$restart-1} {
set X  50.00
set Y  50.00
set Z  50.00
set C  0.00
set C1 0.00
cellBasisVector1    $X      0.0     0.0     #Angstorm
cellBasisVector2    0.0     $Y      0.0     #Angstorm
cellBasisVector3    0.0     0.0     $Z      #Angstorm
cellOrigin    $C    $C1     $C1     #Angstorm
```

Explanation:

if {$restart-1}: Conditional block for periodic boundary conditions.

set X, Y, Z: Sets the dimensions of the simulation box.

cellBasisVectorl, 2, 3: Defines the basis vectors of the simulation cell.

cellOrigin: Sets the origin of the cell.

```
# PME (for full-system periodic electrostatics)
PME                  yes
PMEGridSpacing       1.0
PMETolerance         0.000001    ;# The accuracy of computing
the Ewald real space (direct) term
wrapAll              on
```

Explanation:

PME: Particle mesh Ewald (PME) calculations

PMEGridSpacing

PMEGridSizeX, Y, Z: Sets the grid size for PME calculations.

wrapAll: Ensures all atoms are wrapped into the primary simulation cell.

```
# Constant Temperature Control
if {0} {
langevin             on    ;# do langevin dynamics
langevinDamping      1     ;# damping coefficient (gamma) of 1/
ps
langevinTemp         $temperature
langevinHydrogen     off    ;# don't couple langevin bath to
hydrogens
}
```

Explanation:
- if {0}: Conditional block for Langevin dynamics (not active).
- langevin: Enables Langevin dynamics.
- langevinDamping: Sets the damping coefficient.
- langevinTemp: Sets the target temperature for Langevin dynamics.
- langevinHydrogen: Excludes hydrogens from Langevin dynamics.

```
# Constant Pressure Control (variable volume)
if {0} {
useGroupPressureyes ;# needed for rigidBonds
useFlexibleCell       no
useConstantArea       no
langevinPiston        on
langevinPistonTarget  1.01325 ;#  in bar -> 1 atm
langevinPistonPeriod  100.0
langevinPistonDecay   50.0
langevinPistonTemp    $temperature
}
```

Explanation:
- if {0}: Conditional block for constant pressure control (not active).
- useGroupPressure: Uses group pressure control (required for rigid bonds).
- useFlexibleCell: Disables flexible cell (constant volume).
- useConstantArea: Disables constant area.
- langevinPiston: Enables Langevin piston for pressure control.
- langevinPistonTarget: Sets the target pressure (1 atm).
- langevinPistonPeriod: Sets the piston period.
- langevinPistonDecay: Sets the piston decay time.
- langevinPistonTemp: Sets the temperature for the Langevin piston.

```
# Fixed Atoms Constraint (set PDB beta-column to 1)
if {0} {
fixedAtoms           on
fixedAtomsForces     on #Is it necessary?
fixedAtomsFile       myfixedatoms.pdb
fixedAtomsCol        B
}
```

Explanation:

if {0}: Conditional block for fixed atom constraints (not active).

fixedAtoms: Enables fixed atom constraints.

fixedAtomsForces: Enables calculation of forces on fixed atoms.

fixedAtomsFile: Specifies the PDB file with fixed atoms.

fixedAtomsCol: Specifies the column in the PDB file that indicates fixed atoms.

```
# IMD Settings (can view sim in VMD)
if {0} {
IMDon               on
IMDport             3000      ;# port number (enter it in VMD)
IMDfreq             1         ;# send every 1 frame
IMDwait             no        ;# wait for VMD to connect before
running?
}
```

Explanation:

if {0}: Conditional block for interactive molecular dynamics (IMD) settings (not active).

IMDon: Enables IMD.

IMDport: Sets the port number for IMD.

IMDfreq: Sets the frequency of frames sent to IMD.

IMDwait: Specifies whether to wait for VMD connection before running.

```
# Output
outputName               $outputname
XSTfile                  $outputname.xst

restartfreq              500       ;# 500steps = every 1ps
dcdfreq                  2000
outputEnergies           100
outputPressure           100
XSTfreq                  500


binaryoutput  no
binaryrestart no
```

Explanation:

outputName: Sets the prefix for output files.

restartfreq: Frequency of restart file output (every 1000 steps).

dcdfreq: Frequency of DCD trajectory output (every 500 steps).

xstFreq: Frequency of extended system output (every 1000 steps).

outputEnergies: Frequency of energy output (every 1000 steps).

outputPressure: Frequency of pressure output (every 1000 steps).

binaryoutput: Disables binary output.

binaryrestart: Disables binary restart files.

```
#############################################################
## EXTRA PARAMETERS      ##
#############################################################

#############################################################
## EXECUTION SCRIPT      ##
#############################################################

# Minimization
minimization on
seed            12345      ;# Random number seed used to
generate initial Maxwell distribution of velocities (d)
minimize     2000000
```

Explanation:

minimize: Specifies the number of minimization steps (2,000,000 steps).

6.2.3 Heating

Heating slowly elevates the temperature of the molecular system from 0 K to the intended goal temperature, around 298 K. This procedure is executed gradually to prevent the introduction of excessive kinetic energy that may destabilize the system. Regulated heating enables the system to adjust to heightened thermal activity, attaining a dynamic equilibrium at the specified temperature. Here is a sample of the **Heating script** for NAMD simulation.

```
#############################################################
## JOB DESCRIPTION      ##
#############################################################

# Heating of 500 molecules of Acetone and 500 molecules of
Methanol

#############################################################
## ADJUSTABLE PARAMETERS      ##
#############################################################
coordinates          ../01_Minimization/DES_DRUG_min.coor
extendedSystem       ../01_Minimization/DES_DRUG_min.xsc

set temperature    298.15
set outputnameDES_DRUG_heat
set restart      0
```

Explanation:

coordinates../01_Minimization/DES_DRUG_min.coor: Specifies the input coordinate file (.coor) from the minimization step.

extendedSystem../01_Minimization/DES_DRUG_min.xsc: Specifies the input extended system file (.xsc) from the minimization step.

set temperature 298.15: Sets the desired target temperature for the simulation to 298.15 K (25°C).

set outputnameDES_DRUG_heat: Defines the output file prefix for the simulation results as DES_DRUG_heat.

set restart 0: Indicates whether to restart the simulation from previous files. A value of 0 means no restart (fresh start).

```
# Continuing a job from the restart files
if {$restart} {
set inputname       $outputname
Coordinates         ../$inputname.restart.coor
Velocities          ../$inputname.restart.vel  ;# remove the
"temperature" entry if you use this!
extendedSystem          ../$inputname.restart.xsc
}
firsttimestep       0
```

Explanation:

if {$restart} {... }: This block of code handles restarting the simulation if the restart variable is set to true (1). It loads restart files for coordinates, velocities, and an extended system if available.

Coordinates../$inputname.restart.coor: Specifies the coordinate file to restart from.

Velocities../$inputname.restart.vel: Specifies the velocity file to restart from.

extendedSystem../$inputname.restart.xsc: Specifies the extended system file to restart from.

firsttimestep 0: Sets the first timestep of the simulation to 0, indicating that this is the start of the simulation.

```
#############################################################
## SIMULATION PARAMETERS ##
#############################################################

# Input
amber           on
parmfile        ../common/DES_DRUG.prmtop
```

Explanation:

amber on: Activates the AMBER force field.

parmfile../common/DES_DRUG.prmtop: Specifies the parameter file (.prmtop) for the system, which contains the force field parameters (bond, angle, nonbonded, etc.).

```
# Force-Field Parameters
dielectric          1.0    # Value of the dielectric constant
and Any value greater than 1.0 will be experimental
exclude             scaled1-4
nonbondedScaling    1.0
1-4scaling          1.0
cutoff              12.0   #Angstorm
switching           on
```

```
switchdist          10.0   #Angstorm
pairlistdist        14.0   #Angstorm
```

Explanation: Same as of Minimization.

```
# Integrator Parameters
timestep            1.0       ;# 1fs/step
rigidDonds          all       ;# Apply SHAKE algorithm to all
covalent bonds involving hydrogens
rigidTolerance      0.00001   ;# Desired accuracy in
maintaining SHAKEed bond lengths
rigidIterations     100       ;# Maximum number of SHAKE
iterations
nonbondedFreq       1
vdwGeometricSigma   yes
fullElectFrequency  2
stepspercycle       20
pairlistsperCycle   2
```

Explanation: Same as of Minimization.

```
# Periodic Boundary Conditions
if {$restart-1} {
set X  50.00
set Y  50.00
set Z  50.00
set C  0.00
set C1 0.00
cellBasisVector1    $X     0.0    0.0    #Angstorm
cellBasisVector2    0.0    $Y     0.0    #Angstorm
cellBasisVector3    0.0    0.0    $Z     #Angstorm
cellOrigin          $C     $C1    $C1    #Angstorm
```

Explanation: Same as of Minimization.

```
# Constant Forces
constantForce      yes
consForceScaling   1.0
# PME (for full-system periodic electrostatics)
PME                 yes
PMEGridSpacing      1.0
PMETolerance        0.000001
wrapAll on
```

Explanation: Same as of Minimization.

```
# Constant Temperature Control
if {0} {
langevin            on    ;# do langevin dynamics
```

```
langevinDamping     1     ;# damping coefficient (gamma) of 1/
ps
langevinTemp        $temperature
langevinHydrogen    off    ;# don't couple langevin bath to
hydrogens
}
```

Explanation: Same as of Minimization.

```
# Constant Pressure Control (variable volume)
if {0} {
useGroupPressureyes ;# needed for rigidBonds
useFlexibleCell       no
useConstantArea       no
langevinPiston        on
langevinPistonTarget  1.01325 ;#  in bar -> 1 atm
langevinPistonPeriod  100.0
langevinPistonDecay   50.0
langevinPistonTemp    $temperature
}
```

Explanation: Same as of Minimization.

```
# Fixed Atoms Constraint (set PDB beta-column to 1)
if {0} {
fixedAtoms          on
fixedAtomsForces    on #Is it necessary?
fixedAtomsFile      myfixedatoms.pdb
fixedAtomsCol       B
}
```

Explanation: Same as of Minimization.

```
# IMD Settings (can view sim in VMD)
if {0} {
IMDon           on
IMDport         3000     ;# port number (enter it in VMD)
IMDfreq         1        ;# send every 1 frame
IMDwait         no       ;# wait for VMD to connect before
running?
}
```

Explanation: Same as of Minimization.

```
# Output
outputName          $outputname
#XSTfile             $outputname.xst
restartfreq         500      ;# 500steps = every 1ps
dcdfreq             2000
outputEnergies      100
outputPressure      100
```

```
binaryoutput  no
binaryrestart no
```

Explanation: Same as of Minimization.

```
#############################################################
## EXTRA PARAMETERS ##
#############################################################

#############################################################
## EXECUTION SCRIPT ##
#############################################################

seed   1010          # Random number seed used to generate
initial Maxwell distribution of velocities
numsteps             600000        # Number of integration steps
temperature          0             # Initial velocity
distribution is performed to meet this Initial temperature (in K)
reassignFreq         1             # Number of steps between
reassignment of velocities (e)
reassignIncr         0.01          # Increment used to adjust
temperature during temperature reassignment
reassignHold         $temperature       # The value of
temperature to be kept after heating is completed
```

Explanation:

seed 1010: Sets the random number seed (1010) for generating initial Maxwell distribution of velocities.

numsteps 600000: Specifies the total number of integration steps (600,000 steps).

temperature 0: Sets the initial temperature for velocity reassignment (0 K).

reassignFreq 1: Sets the frequency (1 step) for reassignment of velocities.

reassignIncr 0.01: Specifies the increment (0.01) for adjusting temperature during reassignment.

reassignHold $temperature: Specifies the final temperature value to hold after heating is completed.

6.2.4 Equilibration

Equilibration enables the system to attain a stable condition at the designated temperature and pressure. Throughout this phase, the system's attributes, including temperature, pressure, and density, are observed to guarantee their stability and consistency. Equilibration guarantees that the molecular system is adequately prepared for the production phase, devoid of early artifacts and temporary impacts. Here is a sample of the **Equilibration script** for NAMD simulation.

```
#############################################################
## JOB DESCRIPTION ##
#############################################################
```

```
# Equilibration of 500 molecules of Acetone and 500 molecules
of Methanol

#############################################################
## ADJUSTABLE PARAMETERS ##
#############################################################

coordinates          ../02_Heating/DES_DRUG_heat.coor
extendedSystem             ../02_Heating/DES_DRUG_heat.xsc
set temperature    298.15
set outputnameDES_DRUG_equilb
set restart     0
```

Explanation:

coordinates../02_Heating/DES_DRUG_heat.coor: Specifies the input coordinates file from the heating stage for the system.

extendedSystem../02_Heating/DES_DRUG_heat.xsc: Specifies the input extended system file from the heating stage.

set temperature 298.15: Sets the temperature of the system to 298.15 K (25°C).

set outputnameDES_DRUG_equilb: Specifies the output file name prefix for the simulation as DES_DRUG_equilb.

set restart 0: Indicates whether to continue from a previous simulation. If restart is 1, the job restarts from previous files.

```
# Continuing a job from the restart files
if {$restart} {
set inputnameDES_DRUG_equilb
Coordinates          $inputname.restart.coor
Velocities           $inputname.restart.vel  ;# remove the
"temperature" entry if you use this!
extendedSystem           $inputname.restart.xsc
}
firsttimestep      0
```

Explanation:

set inputnameDES_DRUG_equilb: Specifies the input name for restart files.

Coordinates, Velocities, and extendedSystem files: Defines the input files for the restart process.

firsttimestep 0: Sets the starting timestep of the simulation to 0.

```
#############################################################
## SIMULATION PARAMETERS ##
#############################################################

# Input
amber  on
parmfile     ../common/DES_DRUG.prmtop
if {$restart-1} {
temperature $temperature
}
```

Explanation: Same as of Minimization.

```
# Force-Field Parameters
dielectric          1.0    # Value of the dielectric constant
and Any value greater than 1.0 will be experimental
exclude             scaled1-4
nonbondedScaling    1.0
1-4scaling          1.0
cutoff              12.0   #Angstorm
switching           on
switchdist          10.0   #Angstorm
pairlistdist        14.0   #Angstorm
```

Explanation: Same as of Minimization.

```
# Integrator Parameters
timestep            1.0        ;# 1fs/step
rigidBonds          all        ;# Apply SHAKE algorithm to all
covalent bonds involving hydrogens
rigidTolerance      0.00001    ;# Desired accuracy in
maintaining SHAKEed bond lengths
rigidIterations     100        ;# Maximum number of SHAKE
iterations
nonbondedFreq       1
vdwGeometricSigma   yes
fullElectFrequency  2
stepspercycle       20
pairlistsperCycle   2
```

Explanation: Same as of Minimization.

```
# Periodic Boundary Conditions
if {$restart-1} {
set X  50.00
set Y  50.00
set Z  50.00
set C  0.00
set C1 0.00
cellBasisVector1    $X     0.0    0.0    #Angstorm
cellBasisVector2    0.0    $Y     0.0    #Angstorm
cellBasisVector3    0.0    0.0    $Z     #Angstorm
cellOrigin          $C     $C1    $C1    #Angstorm
```

Explanation: Same as of Minimization.

```
# PME (for full-system periodic electrostatics)
PME                 yes
PMEGridSpacing      1.0
PMETolerance        0.000001
wrapAll             on
```

Explanation: Same as of Minimization.

```
# Constant Temperature Control
if {1} {
langevin            on    ;# do langevin dynamics
langevinDamping     1     ;# damping coefficient (gamma) of 1/
ps
langevinTemp        $temperature
langevinHydrogen    off    ;# don't couple langevin bath to
hydrogens
}
```

Explanation: Same as of Minimization.

```
# Constant Pressure Control (variable volume)
if {1} {
useGroupPressureyes ;# needed for rigidBonds
useFlexibleCell       no
useConstantArea       no
langevinPiston        on
langevinPistonTarget  1.01325 ;#  in bar -> 1 atm
langevinPistonPeriod  100.0
langevinPistonDecay   50.0
langevinPistonTemp    $temperature
}
```

Explanation: Same as of Minimization.

```
# Fixed Atoms Constraint (set PDB beta-column to 1)
if {0} {
#fixedAtoms           on
#fixedAtomsForces     on #Is it necessary?
#fixedAtomsFileDES_DRUG_equilb.restart.coor
#fixedAtomsCol        B
}
```

Explanation: Same as of Minimization.

```
# IMD Settings (can view sim in VMD)
if {0} {
IMDon           on
IMDport         3000    ;# port number (enter it in VMD)
IMDfreq         1       ;# send every 1 frame
IMDwait         no      ;# wait for VMD to connect before
running?
}
```

Explanation: Same as of Minimization.

```
# Output
outputName            $outputname
XSTfile               $outputname.xst
```

```
restartfreq          500     ;# 500 steps = every 20 ps
dcdfreq              5000
outputEnergies       500
outputPressure       500
binaryoutput    no
binaryrestart   no
```

Explanation: Same as of Minimization.

```
#############################################################
## EXTRA PARAMETERS ##
#############################################################

#############################################################
## EXECUTION SCRIPT ##
#############################################################

# Equilibration
seed                      1010       # Random number seed used to
generate initial Maxwell distribution of velocities
numsteps             5000000              # Number of
integration steps
#temperature         300    # Initial velocity distribution is
performed to meet this Initial temperature (in K)
rescaleFreq          1      # Number of steps between rescaling
velocities to a given temperature
rescaleTemp          $temperature # Temperatures to which
velocities are rescaled
```

Explanation: Same as of Minimization.

6.2.5 Production

The production step entails executing the MD simulation to gather data for analysis. This phase often extends over a considerable period, enabling the system to develop and provide substantial insights into its functionality. Throughout manufacturing, the system's trajectories, energies, and more features are documented, forming the foundation for eventual scientific study and interpretation. Here is a sample of the **Production script** for NAMD simulation.

```
#############################################################
## JOB DESCRIPTION ##
#############################################################
# NVT dynamics of 500 molecules of Acetone and Methanol Each
# for time interval of t_start to t_end ns.
#############################################################
## ADJUSTABLE PARAMETERS ##
#############################################################

coordinates DES_DRUG_equilb.coor
extendedSystem       DES_DRUG_equilb.xsc
Velocities DES_DRUG_equilb.vel
```

```
set temperature 298.15
set outputname DES_DRUG_Prod1
set restart  0

# Continuing a job from the restart files
if {$restart} {
set inputname DES_DRUG_Prod1
Coordinates $inputname.restart.coor
Velocities $inputname.restart.vel;# remove the "temperature"
entry if you use this!
extendedSystem      $inputname.restart.xsc
}
firsttimestep 0

#############################################################
## SIMULATION PARAMETERS ##
#############################################################
# Input
amber  on
parmfile      ../../common/DES_DRUG.prmtop
if {0} {
temperature         $temperature
}
```

Explanation: Same as of Equilibration.

```
# Force-Field Parameters
dielectric          1.0    # Value of the dielectric constant
and Any value greater than 1.0 will be experimental
exclude             scaled1-4
nonbondedScaling    1.0
1-4scaling          1.0
cutoff              12.0   #Angstorm
switching           on
switchdist          10.0   #Angstorm
pairlistdist        14.0   #Angstorm
```

Explanation: Same as of Equilibration.

```
# Integrator Parameters
timestep            1.0  ;# 1fs/step
nonbondedFreq       1
vdwGeometricSigma   yes
fullElectFrequency  2
stepspercycle       20
pairlistsperCycle   2
rigidBonds          all         ;# Apply SHAKE algorithm to
all covalent bonds involving hydrogens
rigidTolerance      0.00001     ;# Desired accuracy in
maintaining SHAKEed bond lengths
rigidIterations     100         ;# Maximum number of SHAKE
iterations
```

Explanation: Same as of Equilibration.

```
# Periodic Boundary Conditions
if {0} {
set X  50.00
set Y  50.00
set Z  50.00
set C  0.00
set C1 0.00
cellBasisVector1    $X      0.0     0.0     #Angstorm
cellBasisVector2    0.0     $Y      0.0     #Angstorm
cellBasisVector3    0.0     0.0     $Z      #Angstorm
cellOrigin          $C      $C1     $C1     #Angstorm
```

Explanation: Same as of Equilibration.

```
# PME (for full-system periodic electrostatics)
PME                 yes
PMEGridSpacing      1.0
PMETolerance        0.000001
wrapAll             on
```

Explanation: Same as of Equilibration.

```
# Constant Temperature Control
if {1} {
langevin            on      ;# do langevin dynamics
langevinDamping     1       ;# damping coefficient (gamma) of 1/
ps
langevinTemp        $temperature
langevinHydrogen    off      ;# don't couple langevin bath to
hydrogens
}
```

Explanation: Same as of Equilibration.

```
# Constant Pressure Control (variable volume)
if {0} {
useGroupPressureyes ;# needed for rigidBonds
useFlexibleCell       no
useConstantArea       no
langevinPiston        on
langevinPistonTarget  1.01325 ;#  in bar -> 1 atm
langevinPistonPeriod  100.0
langevinPistonDecay   50.0
langevinPistonTemp    $temperature
}
```

Explanation:

Constant pressure (NPT) is not used here, as the pressure control section is commented out (if {0}).

The Langevin piston settings are also not active.

```
# Fixed Atoms Constraint (set PDB beta-column to 1)
if {0} {
fixedAtoms          on
#fixedAtomsForces   on #Is it necessary?
fixedAtomsFileDES_DRUG_equilb.coor
fixedAtomsCol       B
}
```

Explanation: Same as of Equilibration.

```
# IMD Settings (can view sim in VMD)
if {0} {
IMDon           on
IMDport         3000    ;# port number (enter it in VMD)
IMDfreq         1       ;# send every 1 frame
IMDwait         no      ;# wait for VMD to connect before
running?
}
```

Explanation: Same as of Equilibration.

```
# Output
outputName              $outputname
XSTfile                 $outputname.xst
velDCDfile              $outputname.velDCD
forceDCDfile    $outputname.forceDCD
restartfreq         500     ;# 500steps = every 1ps
dcdfreq             5000
veldcdfreq          5000
forcedcdfreq        5000
outputEnergies      500
outputPressure      500
binaryoutput  no
binaryrestart no
```

Explanation: Same as of Equilibration.

```
#############################################################
## EXTRA PARAMETERS ##
#############################################################
## EXECUTION SCRIPT ##
#############################################################

# Production
seed   3010   ;# Random number seed used to generate initial
Maxwell distribution of velocities
numsteps      10000000      ;# Number of integration steps
```

Explanation: Same as of Equilibration.

This production run is for 10 ns. One can increase the timestep to any value to achieve the desired simulation time.

ACKNOWLEDGMENT

The discussion of the basics of MD simulations was developed for NPTEL and is reported and hosted on (https://nptel.ac.in/courses/103103036).

REFERENCES

1. Allen, M.P. and D.J. Tildesley, *Computer Simulation of Liquids*. Clarendon Press, Oxford Science Publications, 1987, Walton Street, UK.
2. Leach, A.R. *Molecular Modeling: Principles and Applications*, 2nd edition. Prentice Hall, 2001, Harlow, UK.
3. (a) Kalé, L., S. Robert., M. Bhandarkar, et al., NAMD2: Greater scalability for parallel molecular dynamics. *The Journal of Chemical Physics*, 1999. **151**: pp. 283–312. (b) Phillips, J.C., D.J. Hardy, and J.D.C. Maia, et al., Scalable molecular dynamics on CPU and GPU architectures with NAMD. *The Journal of Chemical Physics,* 2020. **153**(4): p. 044130. https://doi.org/10.1063/5.0014475
4. Maitland, C.G., M. Rigby, and B.E. Smith, et al., *Intermolecular Forces: Their Origin and Determination*. Clarendon Press: Oxford, 1981.
5. Gray, G.C. and E.K. Gubbins, *Theory of Molecular Fluids. 1. Fundamentals*. Clarendon Press: Oxford, 1984.
6. Stone, J.A., *The Theory of Intermolecular Forces*. Clarendon Press: Oxford, 1996.
7. Cornell, D.W., P. Cieplak, and I.C. Bayly, et al., 2nd generation force-field for the simulation of proteins, nucleic-acids, and organic molecules. *Journal of the American Chemical Society*, 1995. **117**: pp. 5179–5197.
8. Brooks, R.B., R.E. Bruccoleri, and B.D. Olafson, et al., CHARMM: A program for macromolecular energy, minimization, and dynamics calculations. *Journal of Computational Chemistry*, 1983. **4**: pp. 187–217.
9. Verlet, L., Computer "experiments" on classical fluids. I. Thermodynamical properties of Lennard-Jones molecules. *Physical Review*, 1967. **159**: pp. 98–103.
10. De Leeuw, S.W., J.W. Perrum, and E.R. Smith, Simulation of electrostatic systems in periodic boundary conditions. I. Lattice sums and dielectric constant. *Proceedings of the Royal Society of London. A. Mathematical and Physical Sciences*, 1980. **A373**: pp. 27–56.
11. Heyes, D.M., Electrostatic potential and fields in infinite point charge lattices. *Journal of Chemical Physics*, 1981. **74**: pp. 1924–1929.
12. Tuckerman, M., B.J. Berne, and G.J. Martyna, Reversible multiple time scale molecular dynamics. *Journal of Chemical Physics*, 1992. **97**: pp. 1990–2001.
13. Nosé, S., A molecular dynamics method for simulations in the canonical ensemble. *Molecular Physics*, 1984a. **52**: pp. 255–268.
14. Feller, E.S., Y. Zhang, and W.R. Pastor, et al., Constant pressure molecular dynamics simulation: The Langevin piston method. *Journal of Chemical Physics*, 1995. **103**: pp. 4613–4621.
15. Martyna, J.G., J.D. Tobias, and L.M. Klein, Constant pressure molecular dynamics algorithms. *Journal of Chemical Physics*, 1994. **101**: pp. 4177–4189.
16. Hoover, W.G., Canonical dynamics: equilibrium phase space distributions. *Physical Review A*, 1985. **A31**: pp. 1695–1697.
17. Lifshitz, E.M. and L.D. Landau, *Statistical Physics: Volume 5*. Butterworth: New York, 1985.

18. Prausnitz, J.M., R.N. Lichtenthaler, and E.G. de Azevedo, *Molecular Thermodynamics of Fluid Phase Equilibria*. Prentice Hall: New Jersey, 1999.
19. Frenkel, D. and B. Smit, *Understanding Molecular Simulation: From Algorithms to Applications*, 2nd edition. Academic Press: San Diego, 2002.
20. Kumar, P.P. and S. Yashonath, Structure, conductivity, and ionic motion in $Na_{1+x}Zr_2Si_xP_{3-x}O_{12}$: A simulation study. *The Journal of Physical Chemistry B*, 2002. **106**: pp. 7081–7089.
21. Margulis, J.C., A.H. Stern, and J.B. Berne, Computer simulation of a "green chemistry" room-temperature ionic solvent. *The Journal of Physical Chemistry B*, 2002. **106**: pp. 12017–12021.
22. Lindahl, E., B. Hess, and D. Spoel, GROMACS 3.0: A package for molecular simulation and trajectory analysis. *Journal of Molecular Modeling*, 2001. **7**: pp. 306–317.
23. Brünger, A.T., *X-PLOR, Version 3.1, A System for X-ray Crystallography and NMR*. The Howard Hughes Medical Institute and Department of Molecular Biophysics and Biochemistry, Yale University, New Haven, CT, USA, 1992.
24. Jorgensen, W.L., D.S. Maxwell, and J. TiradoRives. Development and testing of the OPLS all-atom force field on conformational energetics and properties of organic liquids. *Journal of the American Chemical Society*, 1996. **118**: pp. 11225–11236.
25. Bernstein, F.C., T.F. Koetzle, and G.J.B. Williams, et al., The protein data bank: A computer-based archival file for macromolecular structures. *Journal of Molecular Biology*, 1977. **112**: pp. 535–542.
26. Case, D.A., K. Belfon, and I.Y. Ben-Shalom et al., *AMBER 14*. University of California: San Francisco, 2014.
27. Martinez, L., R. Andrade, and E.G. Birgin, et al., PACKMOL: A package for building initial configurations for molecular dynamics simulations. *Journal of Computational Chemistry*, 2009. **30**(13): pp. 2157–2164. https://doi.org/10.1002/jcc.21224.
28. Humphrey, W., A. Dalke, and K. Schulten, VMD: Visual molecular dynamics. *Journal of Molecular Graphics*, 1996. **14**(1): pp. 33–38. https://doi.org/10.1016/0263-7855(96)0 0018-5.
29. Brehm, M. and B. Kirchner, TRAVIS – A free analyzer and visualizer for Monte Carlo and molecular dynamics trajectories. *Journal of Chemical Information and Modeling*, 2011, **51**(8): pp. 2007–2023. https://doi.org/10.1021/ci200217w.

7 Case Study of Molecular Dynamics Simulation

7.1 INTRODUCTION

Two different systems have been selected for the case study of nanoscale molecular dynamics (NAMD) simulation based on their miscibility and phase behavior [1]. Understanding the compatibility of different components in a mixture is mainly determined by the polarity of the molecules. The concept of "like dissolves like" comes into play here, meaning that polar substances blend nicely with other polar substances, while nonpolar substances mix well with other nonpolar ones. Understanding the interactions between molecules is crucial, as it determines their miscibility. Intermolecular forces such as hydrogen bonding, van der Waals forces, and dipole–dipole interactions are particularly important in this regard. Factors such as the size and structure of the molecules can also play a role in determining how well they mix together; molecules that are larger or have different structures may not blend as readily. Understanding the impact of temperature on miscibility is crucial as it plays a vital role in overcoming intermolecular forces. Additionally, in gas–liquid systems, pressure can also contribute to the overall miscibility. In addition, the presence of other components in the mixture can affect the ability of substances to mix, either increasing or decreasing their miscibility.

In the following sections, we have examined various mixtures of polar and nonpolar compounds as model systems for NAMD simulation. These systems are made up of a variety of different components, including acetone, methanol, benzene, decane, dodecane, and water.

Molecules were constructed in GaussView 6 and underwent geometry optimization with Gaussian 16 at the B3LYP/6-31G* theoretical level [2]. The molecules' partial atomic charges were derived using the restricted electrostatic potential (RESP) module [3] of AMBER 14 [4], and the partial charges were subsequently fitted. The generalized Amber force field functional of the ANTECHAMBER module [5] of AMBER 14 was used to create all force field characteristics. Initially, a designated quantity of component molecules was arranged in a simulation box using PACKMOL [6]. Table 7.1 represents the system configuration for all the different systems. Subsequently, the molecular dynamics simulation for each system was conducted using the NAMD 2.10 software at 298.15 K and atmospheric pressure [1].

The NAMD simulation consisted of a 6 ns minimization phase, followed by a 0.5 ns heating period at 298.15 K, and concluded with a 10 ns equilibration phase. The NPT ensemble was used to equilibrate the system, during which the simulation box size was adjusted for equilibrium. A consistent system energy has been achieved to verify the successful conclusion of the equilibration phase and to proceed with the production run. A 50 ns production run using the NVT ensemble at 298.15 K

DOI: 10.1201/9781032694733-7

TABLE 7.1
Molecular Configuration for Different Systems

System	No. of Molecules of Component 1	No. of Molecules of Component 2	System Characteristic
Acetone–methanol	Acetone-500	Methanol-500	Homogeneous
Benzene–decane	Benzene-500	Decane-500	Homogeneous
Benzene–dodecane	Benzene-500	Dodecane-500	Homogeneous
Benzene–water	Benzene-250	Water-250	Heterogeneous

and 1 atm pressure was conducted. The periodic boundary condition (PBC) was implemented to preserve the system's continuity. The TRAVIS software was used to acquire the spatial density function (SDF) data, using the parameter file and the final trajectory file [7]. The VMD 1.9.3 trajectory analyzer visualization tool was used to acquire the SDF isosurfaces and the angular distribution results derived from the MD simulation. Designated references and observed molecules were chosen to generate the SDF coordinate file. Appropriate isovalues were established to exhibit the three-dimensional SDF maps for the systems.

The molecular configurations for the different systems are as follows:

Following are the step-by-step method protocol for loading parameter and trajectory files obtained from the NAMD simulation run. These steps will require a proper order in terms of file selection, browsing, loading, and running specific inbuilt scripts within VMD [7]. Proper visualization of the different trajectory files and the system box is the main focus of this protocol.

Step 1: Open VMD 1.9.3

Step 2: In the VMD Main window: Select **File** → **New Molecule** → **Browse** and select **(.prmtop)** file → Click on **Load** (This will load the forcefield parameter file for the system)

Figure 7.1 shows the original VMD Main window at the left and the Molecule File Browser window at the right.

Step 3: Again click on **Browse** → select **.dcd** file → Click on **Load** → System box will appear

For visualization of the PBC box as applied during the simulation, use the following command in the command window of VMD [7]: **pbc box -center origin -color red** (Figures 7.2–7.4)

Step 4: Click on **Graphics** → Select **Representations** from the dropdown menu → **Graphical Representations** window will appear.

To observe the name of the components within the system as specified during input file preparation (.mol2, .pdb), we go to **Selections** tab and click on **resname.**

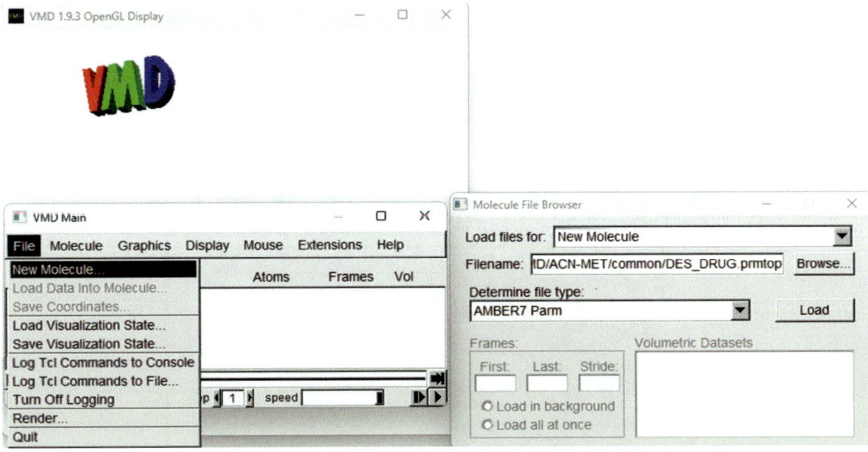

FIGURE 7.1 VMD main window and file loading window.

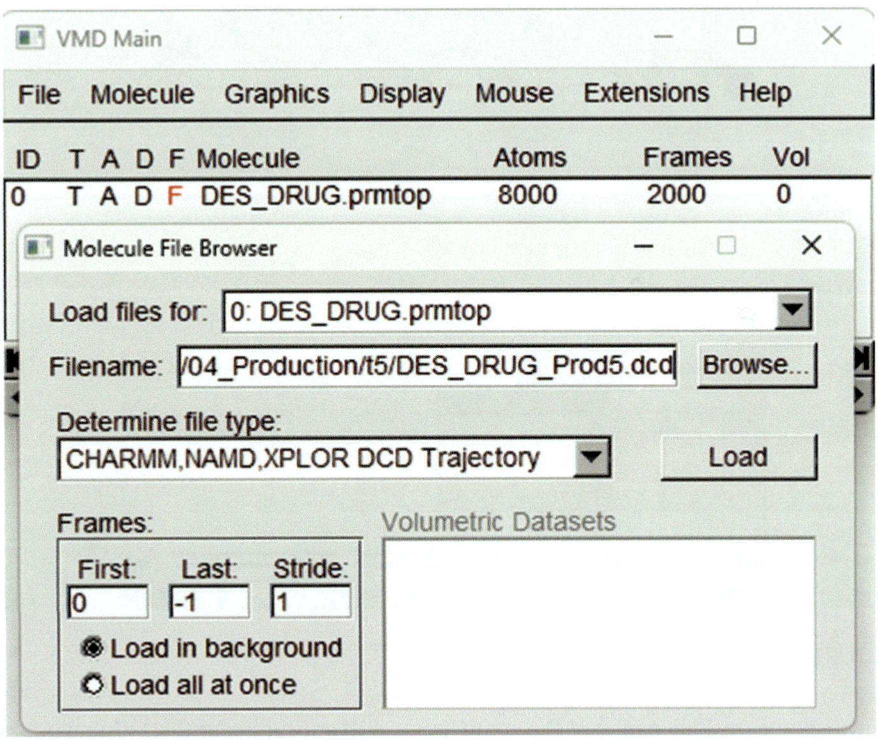

FIGURE 7.2 Parameter file and trajectory file loading.

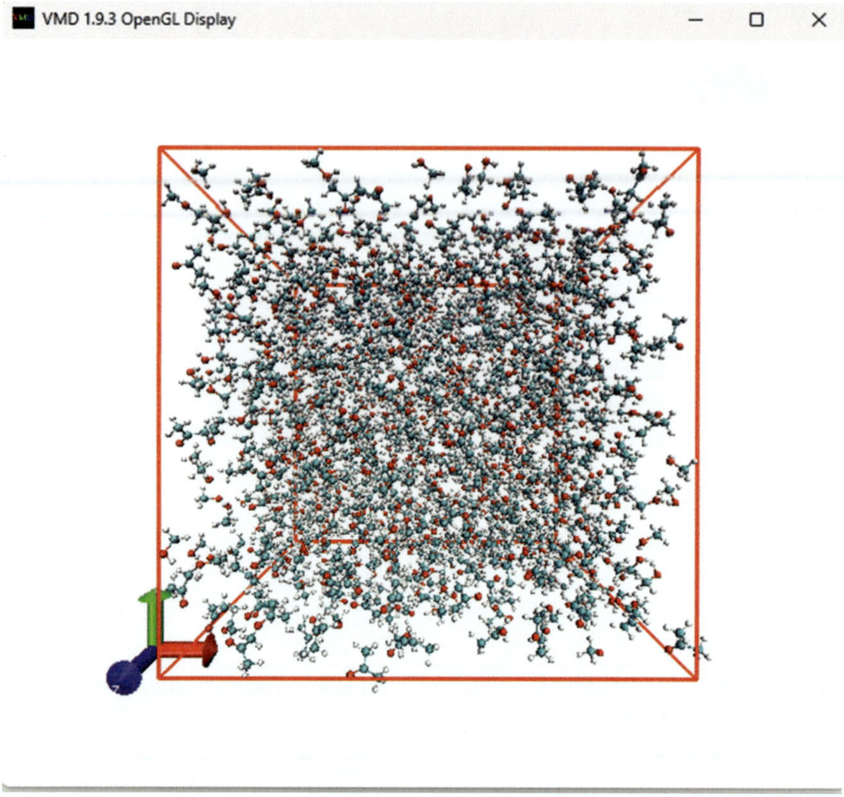

FIGURE 7.3 System box visualization in VMD.

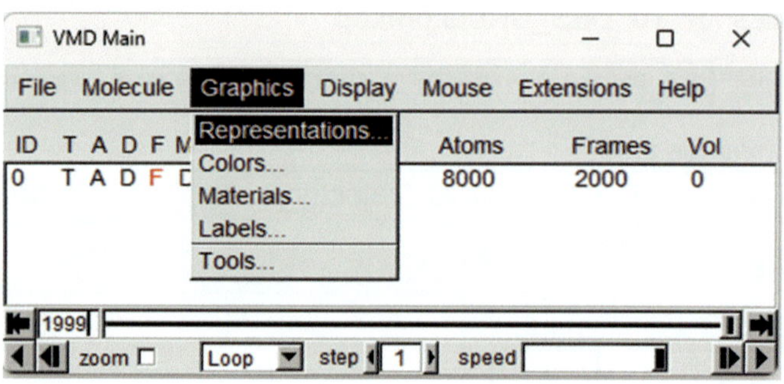

FIGURE 7.4 Graphics window in VMD.

The resname (residue name) of the components will appear under the **Value** window, as can be observed in Figure 7.5.

Step 5: To save .bmp image file, click on **File** → Select **Render** from the drop-down menu

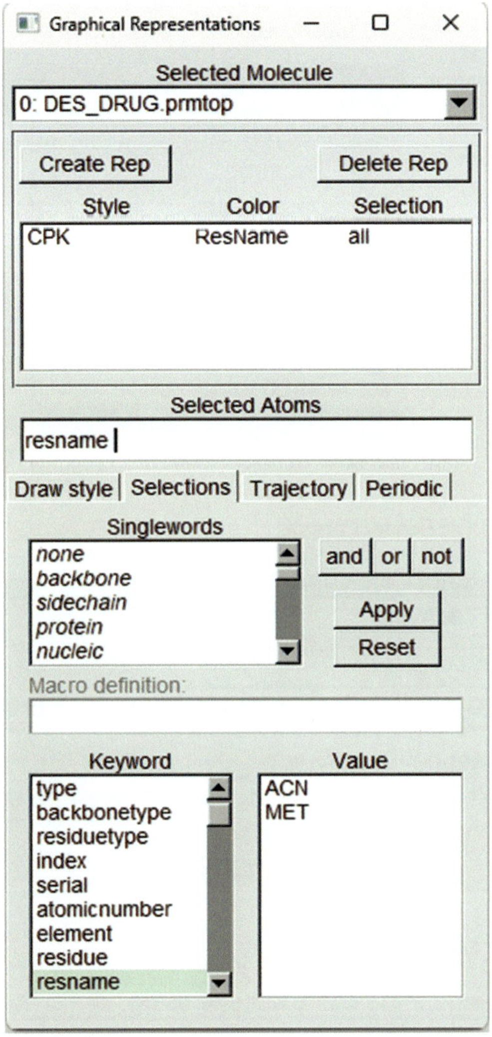

FIGURE 7.5 Graphical representations window in VMD.

File Render Controls window will appear. Select **Tachyon** from the dropdown list → specify Filename by browsing→ keep default settings in **Render Command** → click on **Start Rendering.** The image file (.bmp format) will be saved in the specified location. Figure 7.6b–e show the image file rendering commands for four different systems, namely, acetone–methanol, benzene–decane, benzene–dodecane, and benzene–water.

In the following sections, we explored the post-simulation analysis of all four systems individually starting with the three homogeneous systems, followed by one heterogeneous system.

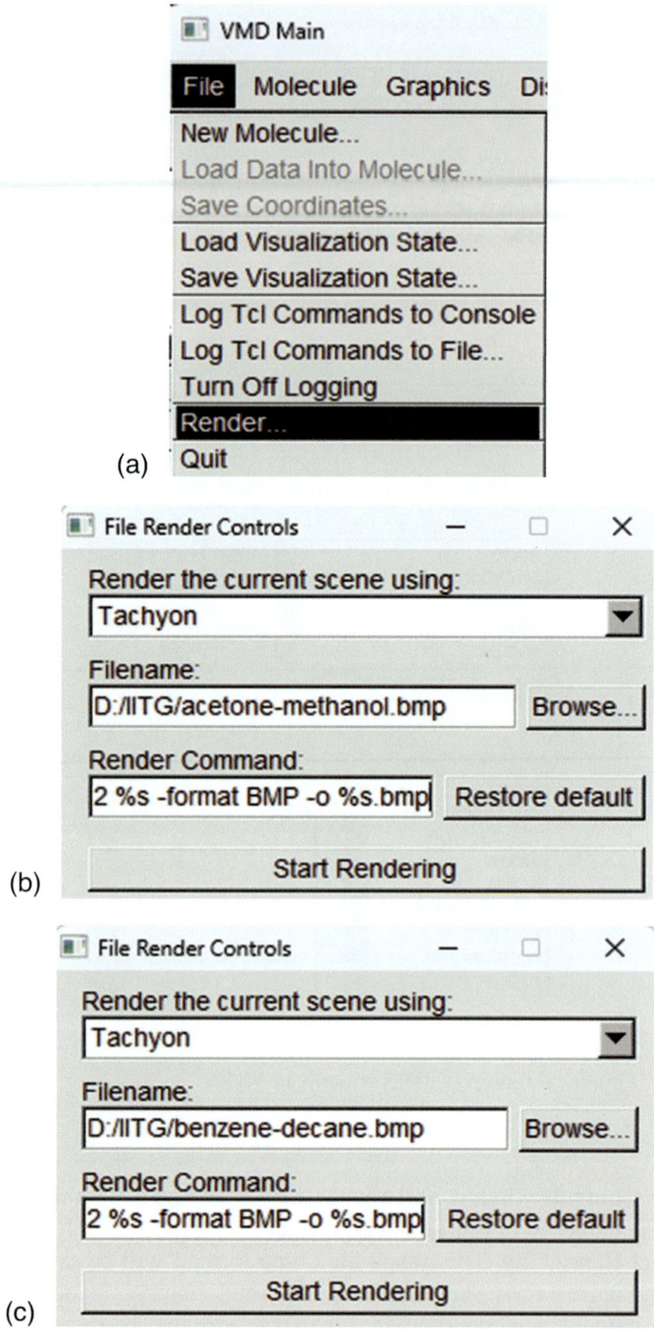

FIGURE 7.6 Image file rendering in VMD. (a) General image file rendering. (b) Image file rendering for acetone-methanol system. (c) Image file rendering for benzene-decane system. (d) Image file rendering for benzene-dodecane system. (e) Image file rendering for benzene-water system.

(*Continued*)

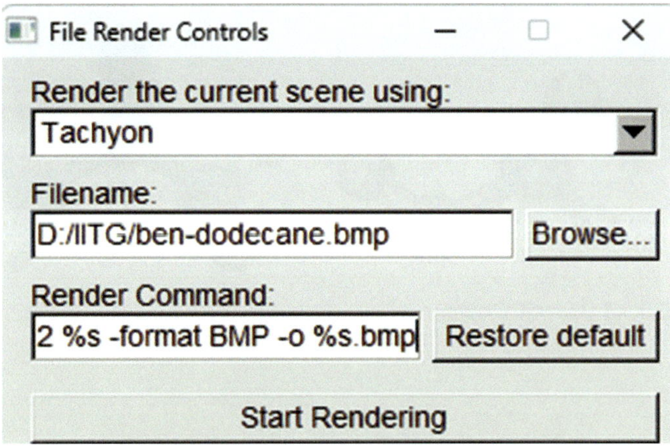

(d)

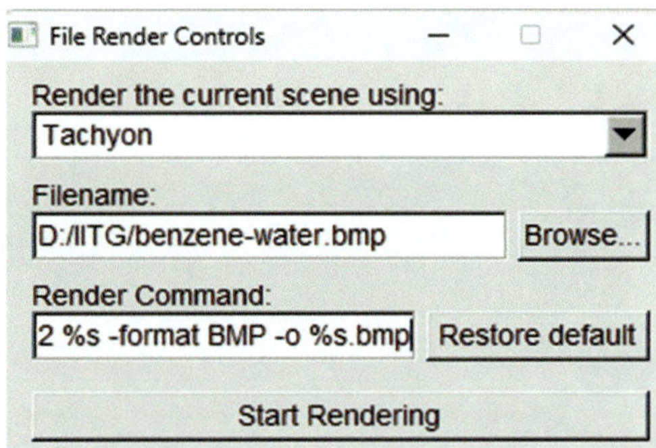

(e)

FIGURE 7.6 (*Continued*) Image file rendering in VMD. (a) General image file rendering. (b) Image file rendering for acetone-methanol system. (c) Image file rendering for benzene-decane system. (d) Image file rendering for benzene-dodecane system. (e) Image file rendering for benzene-water system.

7.2 HOMOGENEOUS SYSTEM

The current section deals with the post-processing and analysis of the homogeneous systems. Structural properties, interaction energy analysis, hydrogen bonding, and diffusive properties will be discussed in this part.

7.2.1 Acetone–Methanol System

Load the parameter file (.prmtop) and the trajectory file (.dcd) for acetone–methanol system in VMD. Adjust the parameters in **Graphical Representations** window setting **resname ACN** and **resname MET.** Select **ColorID** from the dropdown menu of **Coloring Method** tab and **CPK** as the **Drawing Method**. We have assigned **ColorID 1** to represent the color "**red**" for acetone molecules, and **ColorID 0** to represent the color "**blue**" for methanol molecules, as shown in Figure 7.7b. Other parameters, such as sphere scale, sphere resolution, bond scale, and bond resolution, can be optimized as per the required visual representation for the respective systems.

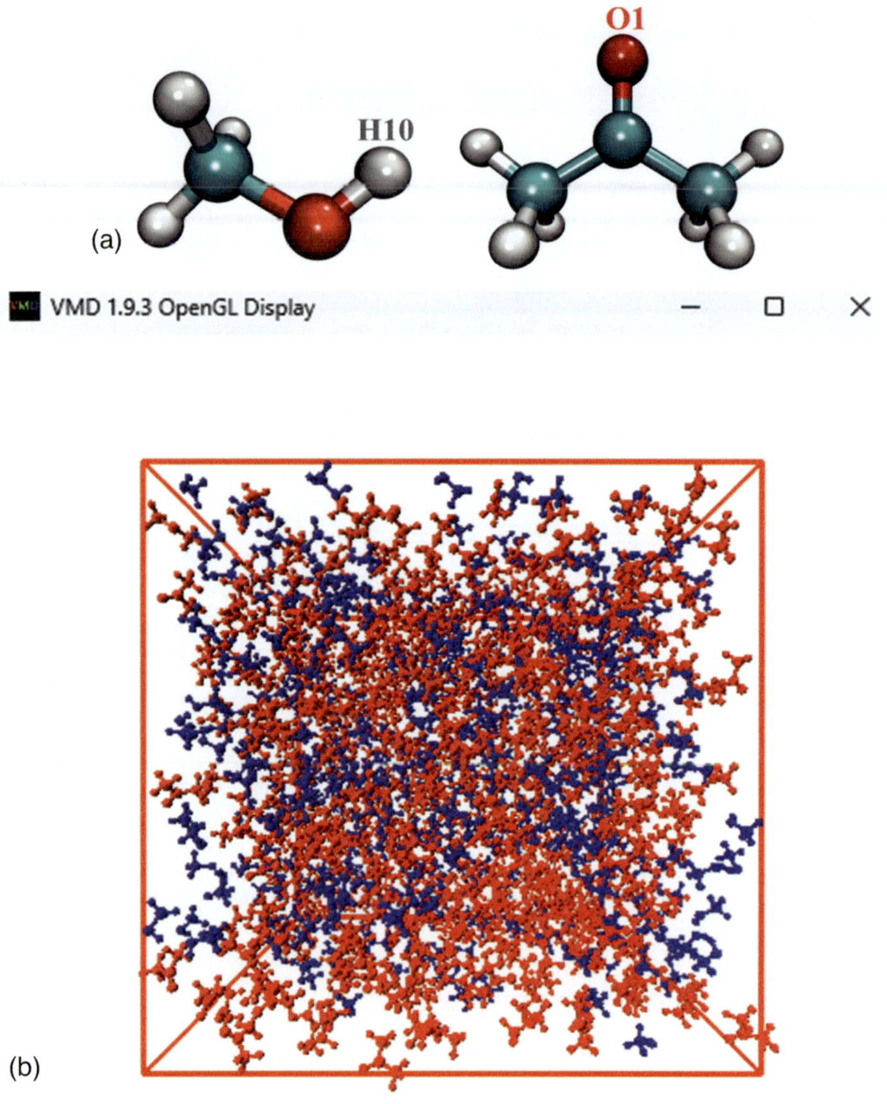

FIGURE 7.7 Configuration of acetone-methanol system. (a) Chemical structures of acetone and methanol. (b) System box visualization (acetone: red, methanol: blue).

7.2.1.1 Structural Properties

Structural properties in NAMD are the characteristics of molecular systems that can be analyzed through molecular dynamics simulations. These properties facilitate understanding the configuration and behavior of molecules over time. In this section, we will focus on two specific structural properties: radial distribution function (RDF) and spatial distribution function (SDF).

7.2.1.1.1 Radial Pair Distribution Function g(r)

Understanding the RDF, also referred to as $g(r)$, is crucial in statistical mechanics as it allows us to analyze how particle density changes with distance from a reference particle. It offers insights into the organization and composition of particles within a system, usually in liquids or amorphous solids. When the RDF is normalized, a value of $g(r)=1$ represents an ideal gas with particles evenly distributed. If $g(r)>1$, it suggests a structured arrangement and a higher probability of finding a particle at a certain distance r. On the other hand, if $g(r)<1$, it indicates a lower probability of finding a particle. Understanding the short-range order in disordered systems is essential, and the RDF plays a vital role.

Click on **Extensions** → Go to **Analysis** → Select **Radial Pair Distribution Function g(r)** from the dropdown list → **Radial Pair Distribution Function g(r)** window will appear, as shown in Figure 7.8a.

In the **Radial Pair Distribution Function g(r)** window, select **.prmtop** file in Molecule, write **name O1** (for acetone) as **Selection 1** and **name H10** (for methanol) as **Selection 2.** Set **max. r.** value as 10.0. This value can be set according to the box dimension (in Å). In general, half of the box size is considered. Tick the boxes of

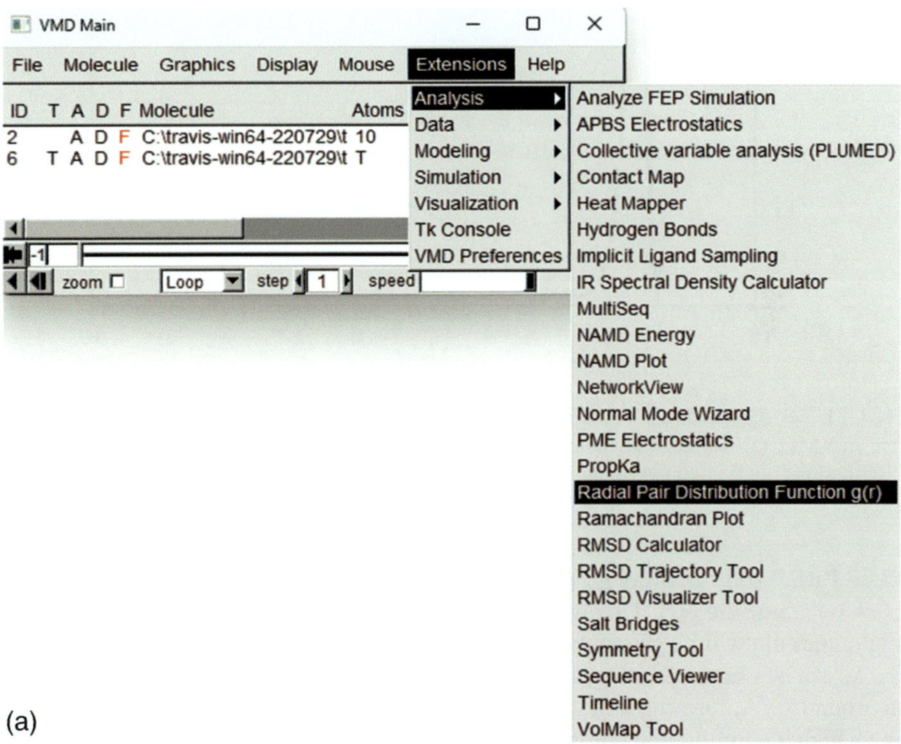

FIGURE 7.8 Radial distribution function (RDF) for acetone-methanol system in VMD. (a) 'Radial Pair Distribution Function g(r)' extension in VMD. (b) Radial pair distribution function g(r) window in VMD. (c) RDF plot for acetone (O1)- methanol (H10).

(Continued)

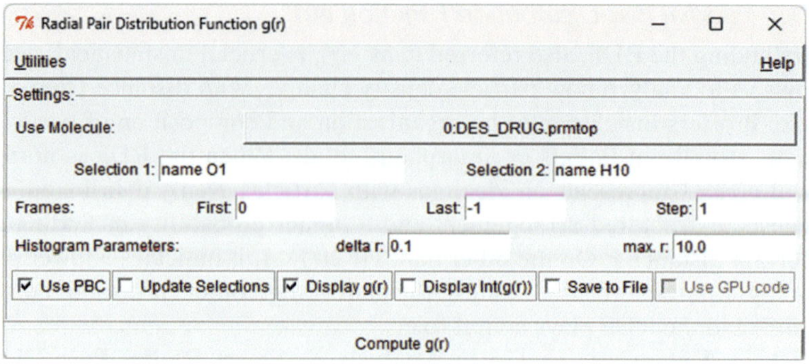

(b)

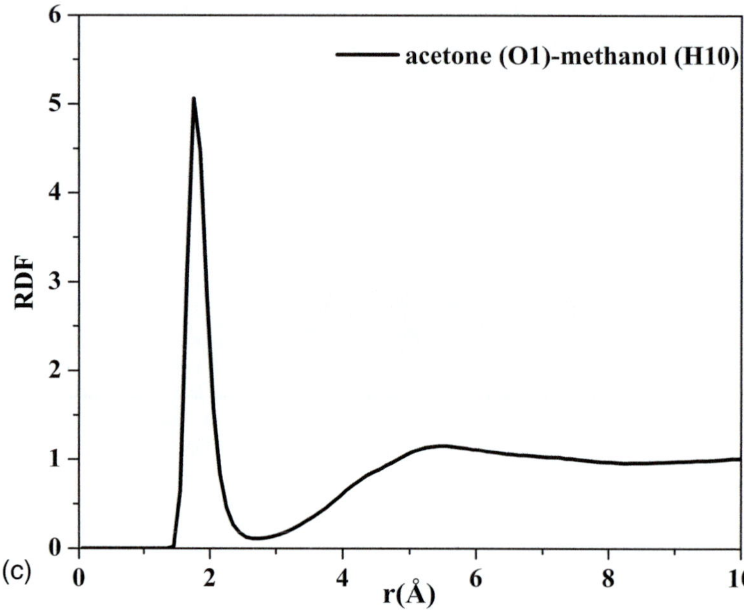

(c)

FIGURE 7.8 (*Continued*) Radial distribution function (RDF) for acetone-methanol system in VMD. (a) 'Radial Pair Distribution Function g(r)' extension in VMD. (b) Radial pair distribution function g(r) window in VMD. (c) RDF plot for acetone (O1)- methanol (H10).

"Use PBC" and **"Display *g(r)*".** All the other values will be kept as default. Then click on **Compute *g(r)*.** This will run the RDF plugin, and after a while, the following output file will be generated:

As can be observed in Figure 7.8c, acetone molecules obtained a higher *g*(*r*) peak at around 1.8 Å, forming a first solvation shell. The *g*(*r*) values peaked to 5, suggesting a higher coordination number. The polar groups present in acetone and methanol form a favorable interactive region through possible O (acetone)….H-O (methanol) hydrogen bonding. A comparatively weak second solvation shell can be observed at around 5.5 Å, suggesting the gradual weakening of the correlation factor among the acetone and methanol molecules.

7.2.1.1.2 Spatial Distribution Function (SDF)

The SDF is a concept that helps us understand how particles or entities are arranged and how dense they are within a specific space. Understanding how particles are distributed relative to each other can reveal interesting patterns, such as clustering, uniform distribution, or dispersion.

Following are the steps involved in saving coordinate files using VMD for SDF generation.

Load **(.prmtop)** file and subsequently the **.dcd** file in VMD → Select the loaded molecule on the **VMD Main** window → Go to **File** → Click on **Save Coordinates** → **Save Trajectory** window will appear → Select **pdb** from the dropdown menu of **File type** → Click on **Save** → Specify a file name for the coordinate (here we have specified **ACM.pdb**) → Coordinate file (**in pdb format**) will be saved at the specified directory. Figure 7.9a and b displays the respective VMD windows regarding the same.

FIGURE 7.9 Spatial distribution function (SDF) for acetone-methanol system in VMD. (a) 'Save coordinates' extension for acetone-methanol system in VMD. (b) Coordinate file saving for acetone-methanol system. (c) Trajectory coordinate file loading for acetone-methanol system in TRAVIS. (d) Input steps for SDF calculation for acetone-methanol system in TRAVIS. (e) Reference molecule of acetone in the VMD display. (f) Isosurface view of reference molecule of acetone in the VMD display. (g) 'Graphical Representations' window for acetone-methanol SDF. (h) SDF plot of methanol around acetone as central molecule in VMD.

(Continued)

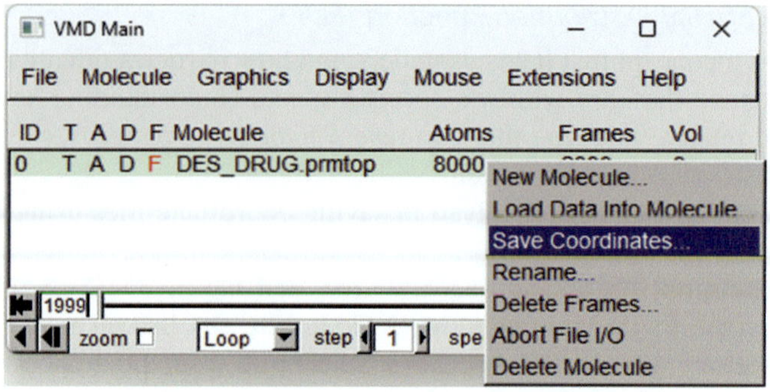

(a)

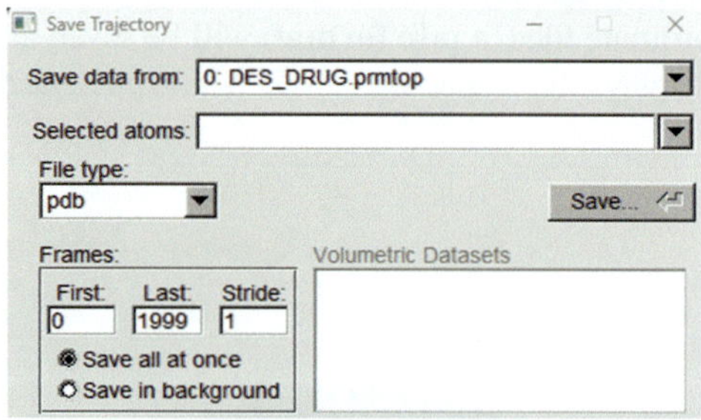

(b)

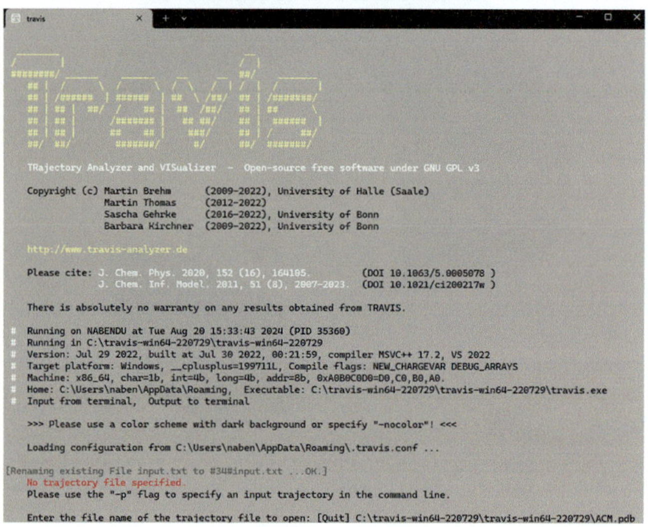

(c)

FIGURE 7.9 (*Continued*) Spatial distribution function (SDF) for acetone-methanol system in VMD. (a) 'Save coordinates' extension for acetone-methanol system in VMD. (b) Coordinate file saving for acetone-methanol system. (c) Trajectory coordinate file loading for acetone-methanol system in TRAVIS. (d) Input steps for SDF calculation for acetone-methanol system in TRAVIS. (e) Reference molecule of acetone in the VMD display. (f) Isosurface view of reference molecule of acetone in the VMD display. (g) 'Graphical Representations' window for acetone-methanol SDF. (h) SDF plot of methanol around acetone as central molecule in VMD.

```
! TRAVIS input file
! Created with TRAVIS version compiled at Jul 30 2022 00:22:24
! Source code version: Jul 29 2022
! Input file written at Wed Aug  7 12:21:06 2024.
C:\travis-win64-220729\travis-win64-220729\ACM.pdb
! Use the advanced mode until the analysis selection menu (y/n)? [no]

! Use these values (y) or enter different values (n)? [yes]

! Update cell geometry in every time step (i.e., NPT ensemble) (y) or use fixed cell (n)? [yes]

! Create images of the structural formulas (y/n)? [no]

! Accept these molecules (y) or change something (n)? [yes]

! Which functions to compute (comma separated)?
SDF
! Use the advanced mode for the main part (y/n)? [no]

! Which of the molecules should be the reference molecule (C3H6O=1, CH4O=2)?
1
! Please enter three comma-separated reference atoms (e.g. "C1,H2,O1"):
C1,H1,O1
! Perform this observation intramolecular (within the reference molecule) (0) or intermolecular (1)? [1]

! Which molecule should be observed (C3H6O=1, CH4O=2)?
2
! Which atoms to observe (e.g. "C1,C3-5,H", "*"=all)? [#2]
*
! Please enter maximal observation radius of this SDF (in pm): [1000.0]
1500
! Please enter binning resolution of this SDF per dimension: [100]

! Add a condition to this observation (y/n)? [no]

! Add another observation (y/n)? [no]

! Map a quantity to the surface of the SDF (y/n)? [no]

! Up to which degree should the SDFs be smoothened (0=not at all)? [2]

! Calculate SDF values in nm^-3 (n) or relative to uniform density (y)? [no]

! In which trajectory frame to start processing the trajectory? [1]

! How many trajectory frames to read (from this position on)? [all]

! Use every n-th read trajectory frame for the analysis: [1]
```

(d)

FIGURE 7.9 (*Continued*) Spatial distribution function (SDF) for acetone-methanol system in VMD. (a) 'Save coordinates' extension for acetone-methanol system in VMD. (b) Coordinate file saving for acetone-methanol system. (c) Trajectory coordinate file loading for acetone-methanol system in TRAVIS. (d) Input steps for SDF calculation for acetone-methanol system in TRAVIS. (e) Reference molecule of acetone in the VMD display. (f) Isosurface view of reference molecule of acetone in the VMD display. (g) 'Graphical Representations' window for acetone-methanol SDF. (h) SDF plot of methanol around acetone as central molecule in VMD.

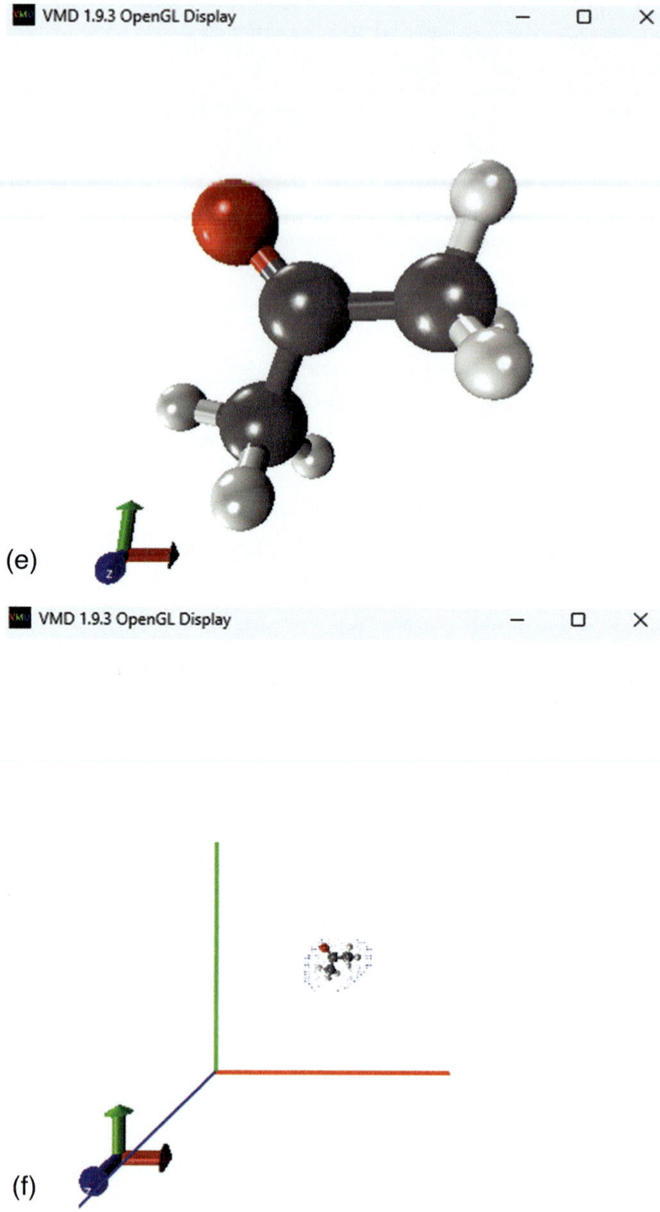

FIGURE 7.9 (*Continued*) Spatial distribution function (SDF) for acetone-methanol system in VMD. (a) 'Save coordinates' extension for acetone-methanol system in VMD. (b) Coordinate file saving for acetone-methanol system. (c) Trajectory coordinate file loading for acetone-methanol system in TRAVIS. (d) Input steps for SDF calculation for acetone-methanol system in TRAVIS. (e) Reference molecule of acetone in the VMD display. (f) Isosurface view of reference molecule of acetone in the VMD display. (g) 'Graphical Representations' window for acetone-methanol SDF. (h) SDF plot of methanol around acetone as central molecule in VMD.

(g)

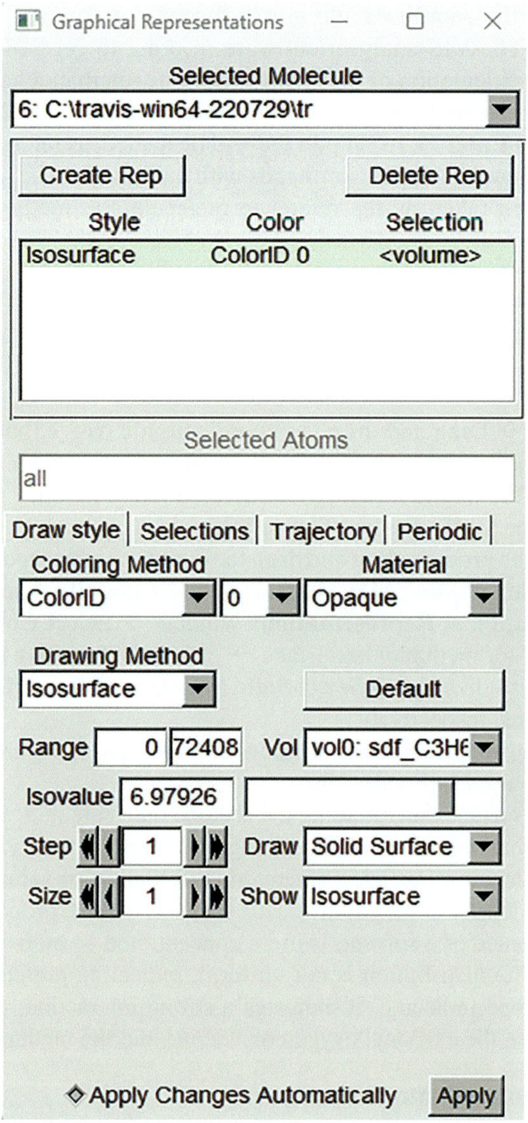

FIGURE 7.9 (*Continued*) Spatial distribution function (SDF) for acetone-methanol system in VMD. (a) 'Save coordinates' extension for acetone-methanol system in VMD. (b) Coordinate file saving for acetone-methanol system. (c) Trajectory coordinate file loading for acetone-methanol system in TRAVIS. (d) Input steps for SDF calculation for acetone-methanol system in TRAVIS. (e) Reference molecule of acetone in the VMD display. (f) Isosurface view of reference molecule of acetone in the VMD display. (g) 'Graphical Representations' window for acetone-methanol SDF. (h) SDF plot of methanol around acetone as central molecule in VMD.

After saving the coordinate file in pdb format, we continue to do more calculations using the TRAVIS analyzer software in order to get SDF plots [8]. Here are the steps for the calculation of SDF for the acetone–methanol homogeneous system.

Open **travis.exe** → Drag and drop the coordinate file (ACM.pdb) to the travis window → Press **Enter** → ACM.pdb file will be loaded, as referenced by Figure 7.9c.

Figure 7.9d shows the input commands with input steps to be followed in TRAVIS. Acetone has been taken as the reference molecule and methanol as the observed molecule.

After the completion of these steps, the Travis window will automatically close and a reference coordinate file (**ref_c3h6o_c1_h1_o1.xyz**) will be created in the Travis directory. Along with this, PLT file (.plt) and Cube file (.cube) will be generated. Using these files, we can obtain an SDF plot through VMD. Here are the steps involved in this.

Open VMD → Drag and drop the coordinate file (**ref_c3h6o_c1_h1_o1.xyz**) to VMD Main window → Open **Graphical Representations** window from **Graphics** tab → Select **CPK** as the **Drawing Method** → The reference molecule of acetone will appear in the VMD display (Figure 7.9e)

To continue the process, drag and drop the coordinate file Open VMD → Drag and drop the PLT file (.plt) to **VMD Main** window → Initially it appears as Figure 7.9f →Open the **Graphical Representations** window → Select **ColorID 0** to represent the color "**blue**" for methanol isosurface → Adjust the **Isovalue** using the adjustment bar for proper visualization → Select **Solid Surface** and **Isosurface** to specify **Draw** and **Show** options, respectively.

The graphical representations window is presented in Figure 7.9g, and the final SDF plot is displayed in Figure 7.9h.

The blue iso-surfaces surrounding the acetone molecule indicate the probable locations of the methanol molecules. These regions suggest areas with a high likelihood of methanol molecules. Surfaces generated with higher iso values generally indicate areas of greater density or probability. This suggests that the property being measured, such as the presence of a solvent, is more concentrated or more likely to be found in these regions. The distribution is not uniform, indicating preferential solvation sites around the acetone molecule. It indicates a strong interaction, potentially hydrogen bonding, between the carbonyl oxygen of acetone and the methanol molecules.

7.2.1.2 Nonbonded Interaction Energy

For nonbonded interaction energy calculation in VMD, the following steps are followed:

Click on **Extensions** → Go to **Analysis** → Select **NAMD Energy** from the dropdown list → The **NAMD Energy** window will appear.

In **NAMD Energy** window, select **.prmtop** file in Molecule, write **resname ACN** (for acetone) as **Selection 1** and **resname MET** (for methanol) as **Selection 2.** Tick the boxes of **VDW, Elec, and Nonbond** that, respectively, indicate van der Waals, electrostatic, and total nonbonded interactions. Set **Output File** name (e.g., ACN-MET-IE). Select the parameter file and XSC file by browsing. All the other values will be kept as the default. Then click on **Run NAMD Energy.** This will run the NAMD energy plugin, and after a while, the following output (Figure 7.10b) file will be generated:

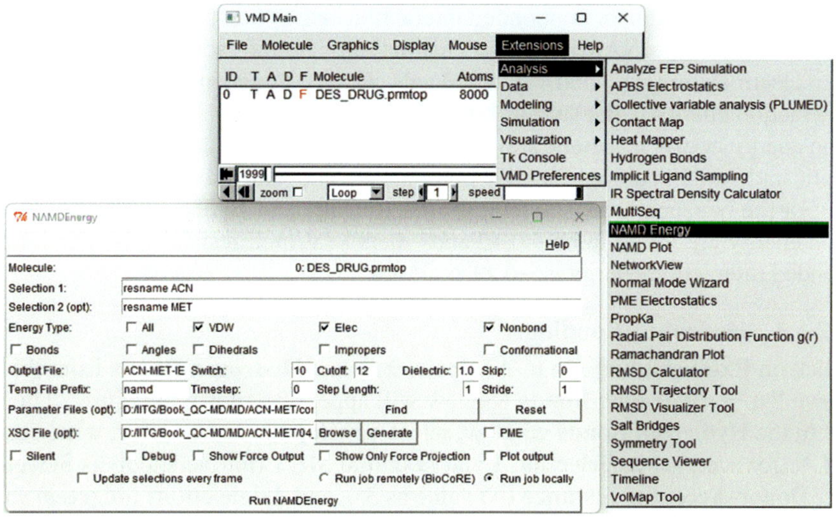

(a)

ACN-MET-IE

Frame	Time	Elec	VdW	Nonbond	Total
0	0	-2399.85	-1485.1	-3884.95	-3884.95
1	1	-2288.18	-1524.94	-3813.12	-3813.12
2	2	-2426.26	-1491.27	-3917.53	-3917.53
3	3	-2416.24	-1515.86	-3932.1	-3932.1
4	4	-2341.09	-1555.86	-3896.95	-3896.95
5	5	-2405.69	-1507.66	-3913.36	-3913.36
6	6	-2447.92	-1554.61	-4002.53	-4002.53
7	7	-2493.23	-1545.88	-4039.11	-4039.11
8	8	-2528.95	-1474.72	-4003.67	-4003.67
9	9	-2492.92	-1501.89	-3994.81	-3994.81
10	10	-2422.27	-1481.14	-3903.41	-3903.41
11	11	-2496.83	-1503.36	-4000.19	-4000.19
12	12	-2432.91	-1577	-4009.91	-4009.91
13	13	-2426.83	-1509.93	-3936.76	-3936.76
14	14	-2452.55	-1541.12	-3993.67	-3993.67
15	15	-2526.18	-1529.56	-4055.74	-4055.74
16	16	-2412.16	-1540.82	-3952.97	-3952.97
17	17	-2310.71	-1564.1	-3874.81	-3874.81
18	18	-2422.1	-1501.64	-3923.74	-3923.74
19	19	-2408.45	-1541.99	-3950.45	-3950.45
20	20	-2490.74	-1498.61	-3989.35	-3989.35
21	21	-2462	-1538.82	-4000.82	-4000.82
22	22	-2463.04	-1530	-3993.04	-3993.04
23	23	-2359.05	-1522.51	-3881.57	-3881.57
24	24	-2462.3	-1482.22	-3944.52	-3944.52
25	25	-2520.2	-1525.41	-4045.61	-4045.61
26	26	-2489.72	-1521.6	-4011.32	-4011.32
27	27	-2402.82	-1536.26	-3939.08	-3939.08
28	28	-2459.48	-1557.6	-4017.08	-4017.08
29	29	-2445.26	-1596.36	-4041.62	-4041.62
30	30	-2326.05	-1488.16	-3814.21	-3814.21

(b)

FIGURE 7.10 Non-bonded interaction energy (NAMD energy) protocol in VMD. (a) NAMD energy inputs for acetone-methanol system. (b) NAMD energy output for acetone-methanol system.

The average value of nonbonded interaction energy can be obtained from the data obtained from the NAMD energy calculation. The interaction energy is classified into electrostatic (E_{elec}) and van der Waals (E_{vdW}) components. The total nonbonded interaction energy is the sum of both of these components. The nonbonded interaction energy between acetone and methanol is found to be –2.40 kcal/mol for electrostatic and –1.52 kcal/mol for vdW component. The total energy is –3.92 kcal/mol.

For the benzene–water system, the electrostatic and vdW components of interaction energy are found to be –0.40 and –0.39 kcal/mol, respectively. The total nonbonded interaction energy is –0.79 kcal/mol.

7.2.1.3 Hydrogen Bonding

Click on **Extensions** → Go to **Analysis** → Select **Hydrogen Bonds** from the dropdown list → **Hydrogen Bonds** window will appear, as shown in Figure 7.11a.

In the **Hydrogen Bonds** window, select **.prmtop** file in Molecule, write **resname ACN** (for acetone) as **Selection 1** and **resname MET** (for methanol) as **Selection 2.** Set **Donor-Acceptor distance** (Å) value as 3.0, and **Angle cutoff (degrees)** value as 180. These are the general criteria for hydrogen bonding. Tick the boxes of **Plot the**

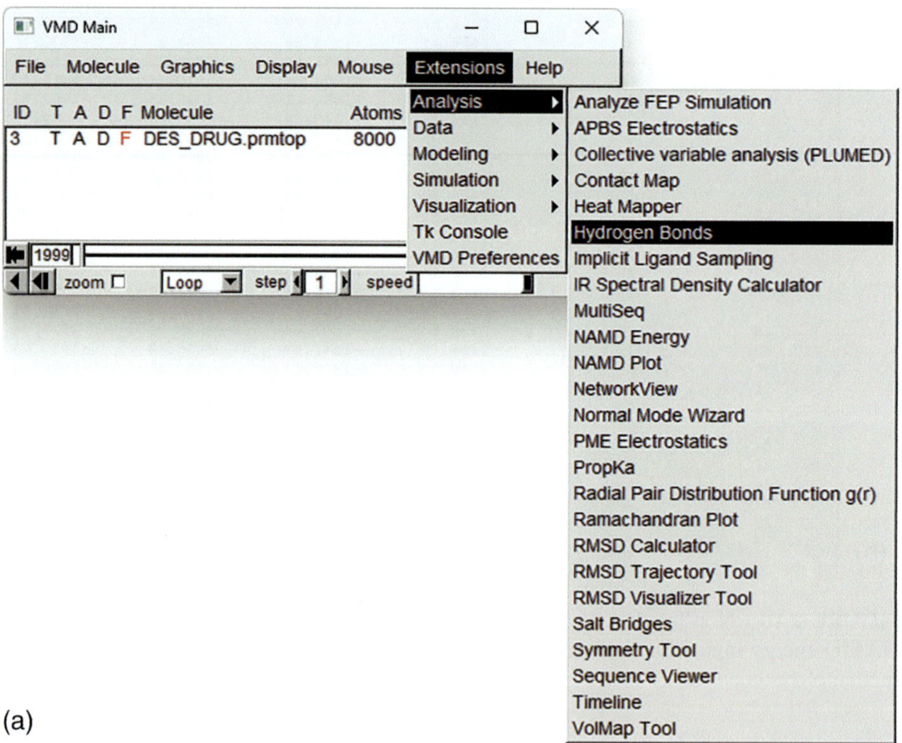

(a)

FIGURE 7.11 Hydrogen bonds protocol for acetone-methanol system in VMD. (a) 'Hydrogen Bonds' extension for acetone-methanol system in VMD. (b) Hydrogen bond inputs for acetone and methanol. (c) Hydrogen bonds output for acetone-methanol system.

(Continued)

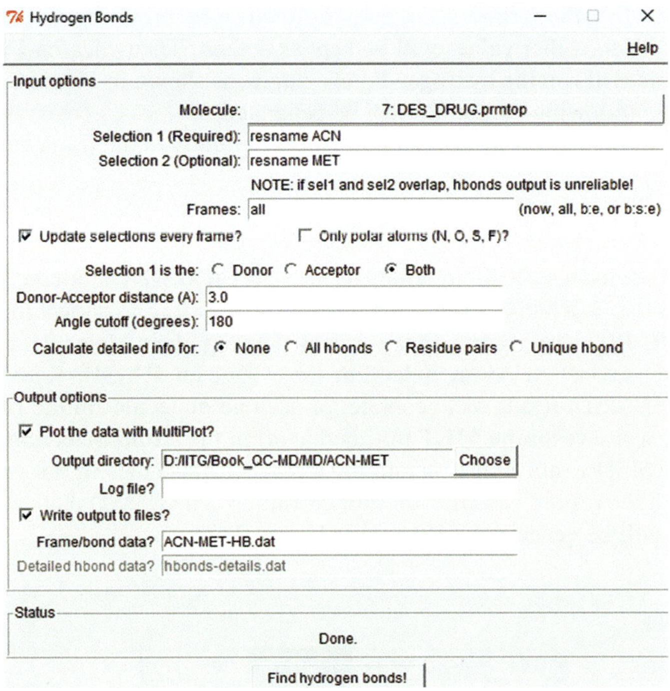

(b)

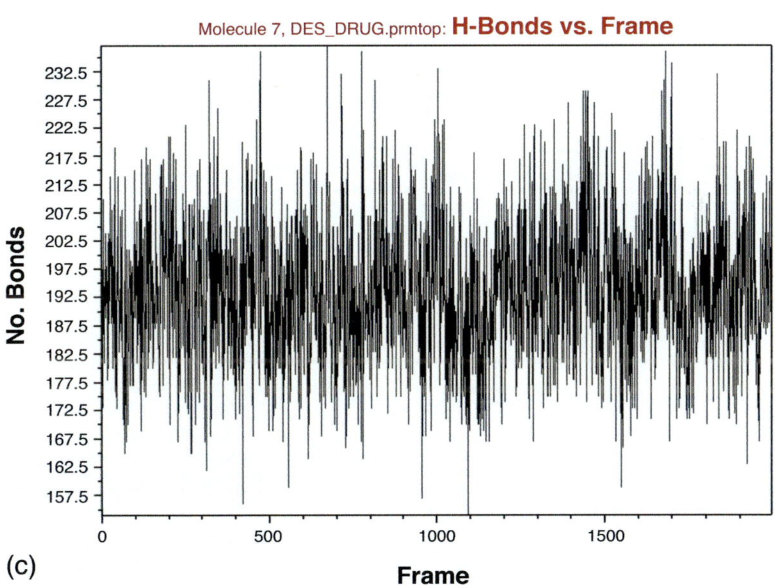

(c)

FIGURE 7.11 (*Continued*) Hydrogen bonds protocol for acetone-methanol system in VMD. (a) 'Hydrogen Bonds' extension for acetone-methanol system in VMD. (b) Hydrogen bond inputs for acetone and methanol. (c) Hydrogen bonds output for acetone-methanol system.

data with MultiPlot, choose the output directory by browsing. Specify the output filename. All the other values will be kept as default. Then click on **Find hydrogen bonds**. This will run the hydrogen bonds plugin, as shown in Figure 7.11b, and after a while, the following output file will be generated:

Figure 7.11c displays the number of hydrogen bonds vs the frame number plot.

7.2.1.4 Dynamic Property

7.2.1.4.1 Root Mean Square Deviation (RMSD)

Click on **Extensions** → Go to **Analysis** → Select **RMSD Visualizer Tool** from the dropdown list → **RMSD Visualizer** window will appear, as shown in Figure 7.12b.

In the **RMSD Visualizer** window, select **.prmtop** file in Molecule, write **resname ACN** (for acetone) in **Atom Selection** tab. Click on **RMSD.** Root mean square deviation (RMSD) result will generate for acetone molecules in the **result** window. Similarly, **write resname MET** (for methanol) in the **Atom Selection** tab. Click on **RMSD.** RMSD result will generate for methanol molecules in the **result** window. Select both the results and click on **Plot result** tab. The RMSD plots for acetone and methanol will be generated, as shown in Figure 7.12c.

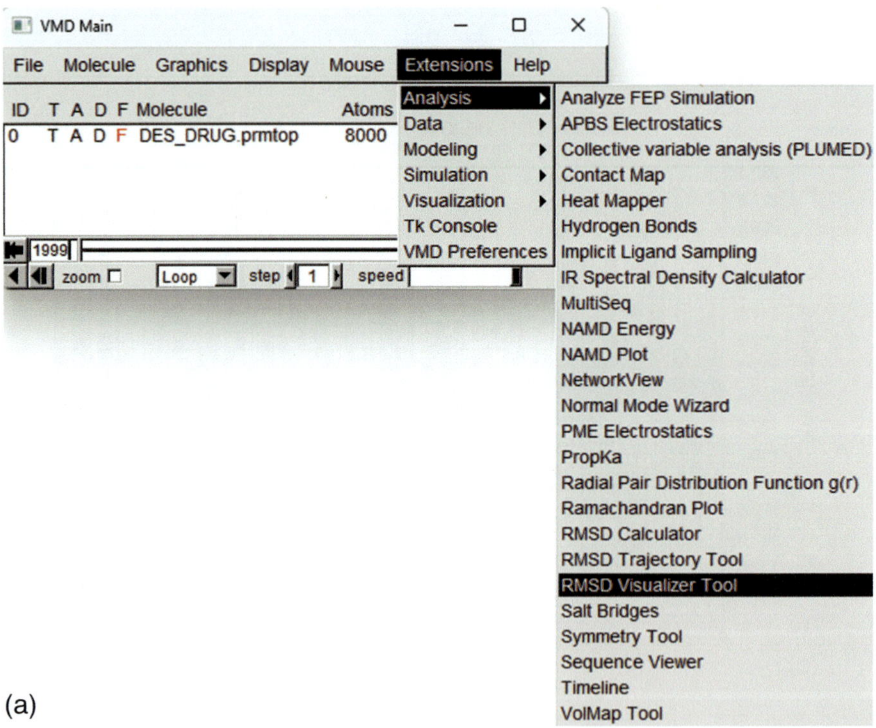

(a)

FIGURE 7.12 Root Mean Square Deviation (RMSD) protocol for acetone-methanol system in VMD. (a) 'RMSD Visualizer Tool' extension in VMD. (b) RMSD visualizer window for acetone-methanol system in VMD. (c) RMSD plot for acetone (blue)-methanol (red) system.

(Continued)

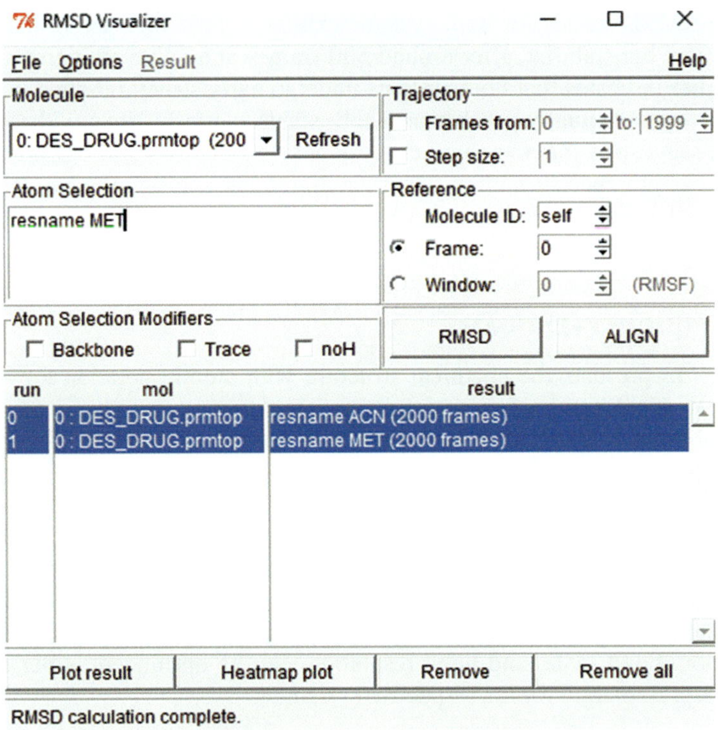

(b)

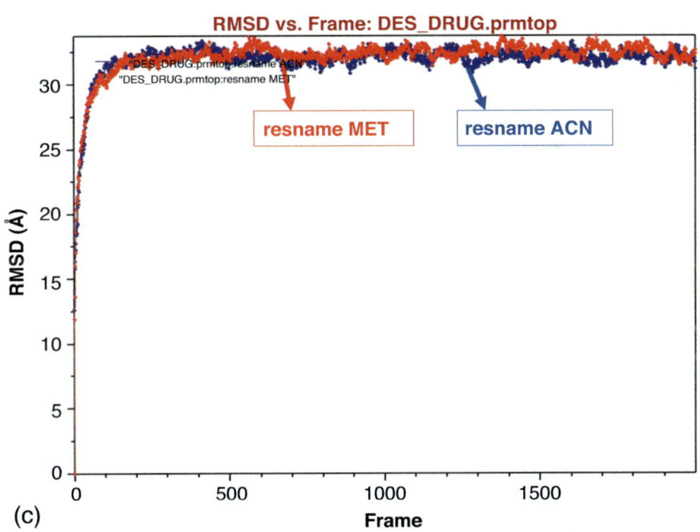

(c)

FIGURE 7.12 (*Continued*) Root Mean Square Deviation (RMSD) protocol for acetone-methanol system in VMD. (a) 'RMSD Visualizer Tool' extension in VMD. (b) RMSD visualizer window for acetone-methanol system in VMD. (c) RMSD plot for acetone (blue)-methanol (red) system.

The RMSD values for both systems exhibit a rapid increase at the start of the simulation, but stabilize after around 500 frames at a value of approximately 30 Å. The figure indicates that both systems undergo early structural alterations or equilibration, but eventually reach reasonably stable conformations with equal RMSD values, suggesting that they have equivalent levels of structural deviation from their respective reference structures. The tight alignment between the two RMSD curves suggests that the two systems have comparable dynamic behavior throughout the experiment.

7.2.2 (Benzene+Decane)

Figure 7.13 presents the chemical structure with atomic notation for benzene and decane molecules.

Load the parameter file (.prmtop) and the trajectory file (.dcd) for the benzene–decane system in VMD. Adjust the parameters in **Graphical Representations** window, setting **resname BEN** and **resname DCN,** for benzene and decane, respectively**.** Select **ColorID** from the dropdown menu of the **Coloring Method** tab and **CPK** as the **Drawing Method**. We have assigned **ColorID 1** to represent the color "**red**" for benzene molecules, and **ColorID 0** to represent the color "**blue**" for decane molecules, as shown in Figure 7.14. Other parameters, such as sphere scale, sphere resolution, bond scale, and bond resolution, can be optimized as per the required visual representation for the respective systems.

7.2.2.1 Structural Properties

Structural properties in NAMD are the characteristics of molecular systems that can be analyzed through molecular dynamics simulations. These properties facilitate understanding the configuration and behavior of molecules over time. In this section, we will focus on two specific structural properties: RDF and SDF.

7.2.2.1.1 Radial Pair Distribution Function g(r)

Click on **Extensions** → Go to **Analysis** → Select **Radial Pair Distribution Function *g*(*r*)** from the dropdown list → **Radial Pair Distribution Function *g*(*r*)** window will appear, as shown in Figure 7.15b.

In the **Radial Pair Distribution Function *g*(*r*)** window, select **.prmtop** file in Molecule, write **name C10** (for decane) as **Selection 1** and **name H36** (for benzene) as **Selection 2.** Set **max. r.** value as 10.0. This value can be set according to the box

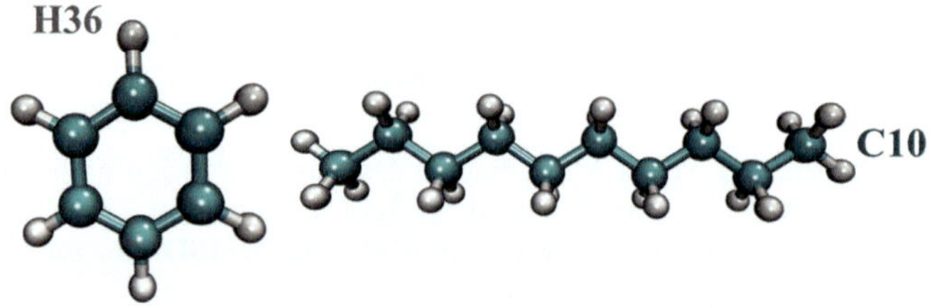

FIGURE 7.13 Chemical structures of benzene and decane.

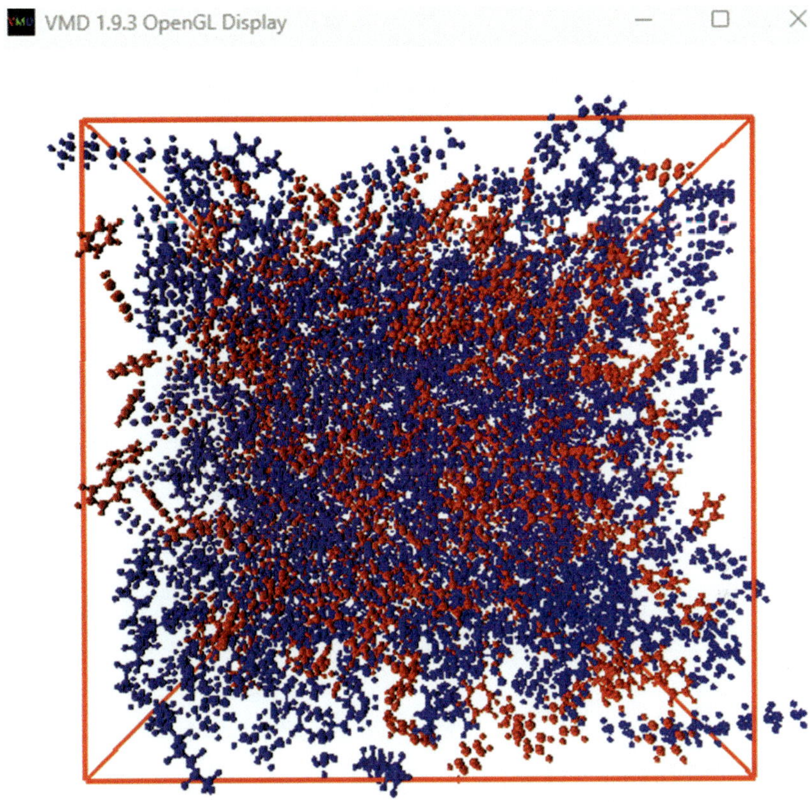

FIGURE 7.14 System box visualization (benzene: red, decane: blue).

dimension (in Å). In general, half of the box size is considered. Tick the boxes of "**Use PBC**" and "**Display *g*(*r*)**". All the other values will be kept as default. Then click on **Compute *g*(*r*)**. This will run the RDF plugin, and after a while, the following output file will be generated:

As can be observed in Figure 7.15c, decane molecules obtained a higher *g*(*r*) peak at around 4.3 Å, forming a first solvation shell. The *g*(*r*) values peaked to 1.1 suggesting a reasonably high coordination number. As both decane and benzene are nonpolar entities, these groups interact by dispersion interaction, and thus a higher *g*(*r*) peak is observed. No further presence of any significant solvation shell is witnessed, suggesting the absence of long-range interactions among the benzene and dodecane molecules, a possible result of the steric effect. It also highlights the fact that the hydrogen bonding interaction is very less in this system.

7.2.2.1.2 Nonbonded Interaction Energy

For nonbonded interaction energy calculation in VMD, the following steps are followed:

Click on **Extensions** → Go to **Analysis** → Select **NAMD Energy** from the dropdown list → The **NAMD Energy** window will appear.

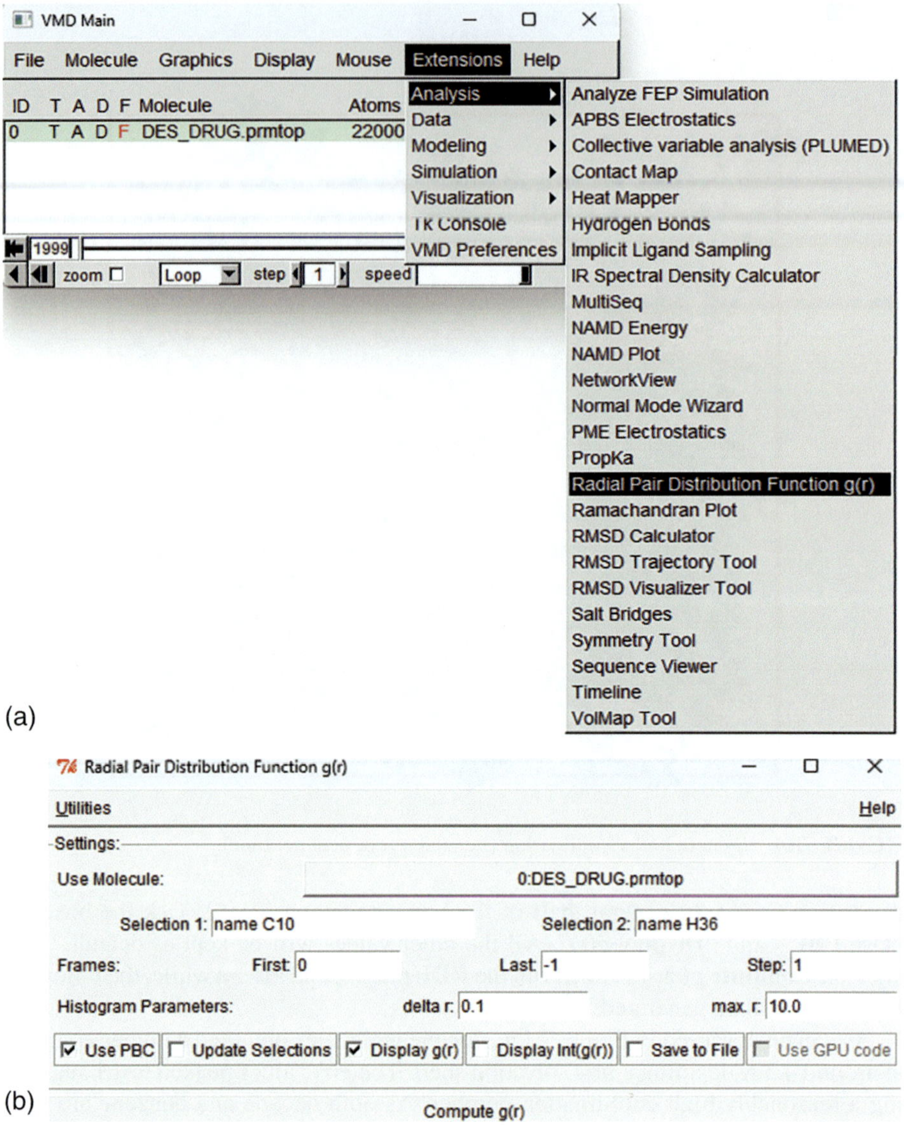

FIGURE 7.15 Radial distribution function (RDF) for benzene-decane system in VMD. (a) 'Radial Pair Distribution Function g(r)' extension for benzene-decane system. (b) Radial pair distribution function inputs for benzene and decane. (c) RDF plot for decane (C10)- benzene (H36).

(Continued)

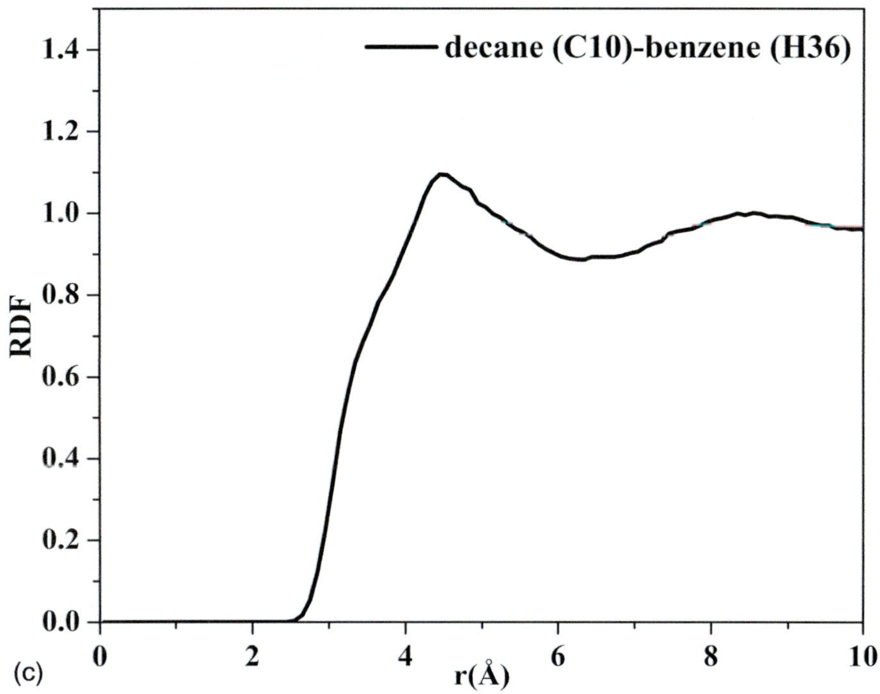

FIGURE 7.15 (*Continued*) Radial distribution function (RDF) for benzene-decane system in VMD. (a) 'Radial Pair Distribution Function g(r)' extension for benzene-decane system. (b) Radial pair distribution function inputs for benzene and decane. (c) RDF plot for decane (C10)- benzene (H36).

In **NAMD Energy** window, select **.prmtop** file in Molecule, write **resname BEN** (for benzene) as **Selection 1** and **resname DCN** (for decane) as **Selection 2.** Tick the boxes of **VDW, Elec, and Nonbond** that respectively indicate van der Waals, electrostatic, and total nonbonded interactions. Set **Output File** name (e.g. BEN-DCN-IE). Select the parameter file and XSC file by browsing. All the other values will be kept as default. Then click on **Run NAMD Energy.** This will run the NAMD energy plugin, and after a while, the following output (Figure 7.16b) file will be generated:

The average value of nonbonded interaction energy can be obtained from the data obtained from the NAMD energy calculation. The interaction energy is classified into electrostatic (E_{elec}) and van der Waals (E_{vdW}) components. The total nonbonded interaction energy is the sum of both of these components.

For the benzene–decane system, the electrostatic and the vdW components of interaction energy are found to be −0.01 and −3.24 kcal/mol, respectively. The total nonbonded interaction energy is −3.25 kcal/mol.

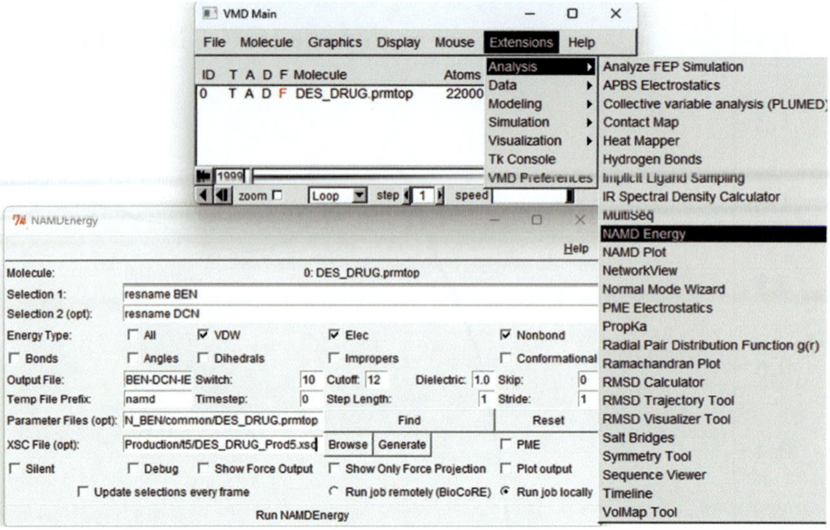

(a)

BEN-DCN-IE

	Frame	Time	Elec	VdW	Nonbond	Total
2	0	0	-7.6345	-3338.48	-3346.11	-3346.11
3	1	1	-13.0873	-3301.81	-3314.9	-3314.9
4	2	2	-6.7223	-3370.86	-3377.58	-3377.58
5	3	3	-6.6744	-3349.57	-3356.24	-3356.24
6	4	4	-5.8245	-3313.78	-3319.61	-3319.61
7	5	5	-11.8653	-3355.36	-3367.23	-3367.23
8	6	6	-8.7383	-3334.75	-3343.48	-3343.48
9	7	7	-10.8496	-3334.82	-3345.67	-3345.67
10	8	8	-4.3278	-3283.19	-3287.52	-3287.52
11	9	9	-10.4341	-3257.95	-3268.38	-3268.38
12	10	10	-10.5784	-3211.6	-3222.18	-3222.18
13	11	11	-8.3217	-3221.14	-3229.47	-3229.47
14	12	12	-10.4864	-3226.93	-3237.42	-3237.42
15	13	13	-10.7371	-3242.37	-3253.11	-3253.11
16	14	14	-15.0768	-3239.66	-3254.73	-3254.73
17	15	15	-9.8835	-3220.54	-3230.43	-3230.43
18	16	16	-8.9871	-3239.27	-3248.26	-3248.26
19	17	17	-7.476	-3273.11	-3280.59	-3280.59
20	18	18	-12.4103	-3247.92	-3260.33	-3260.33
21	19	19	-7.7579	-3278.3	-3286.06	-3286.06
22	20	20	-8.4801	-3226.78	-3235.26	-3235.26
23	21	21	-12.6644	-3269.28	-3281.95	-3281.95
24	22	22	-7.8484	-3280.38	-3288.23	-3288.23
25	23	23	-11.0869	-3345.99	-3357.07	-3357.07
26	24	24	-13.9563	-3344.31	-3358.27	-3358.27
27	25	25	-14.4233	-3306.04	-3320.47	-3320.47
28	26	26	-10.086	-3284.73	-3294.81	-3294.81
29	27	27	-10.3448	-3284.9	-3295.25	-3295.25
30	28	28	-12.3933	-3274.36	-3286.75	-3286.75
31	29	29	-5.8055	-3316.85	-3322.66	-3322.66
32	30	30	-9.3074	-3289.66	-3298.97	-3298.97

(b)

FIGURE 7.16 NAMD energy protocol benzene-decane system in VMD. (a) NAMD energy inputs for benzene-decane system. (b) NAMD energy output for benzene-decane system.

7.2.2.1.3 Hydrogen Bonding

Click on **Extensions** → Go to **Analysis** → Select **Hydrogen Bonds** from the drop-down list → **Hydrogen Bonds** window will appear, as shown in Figure 7.17a.

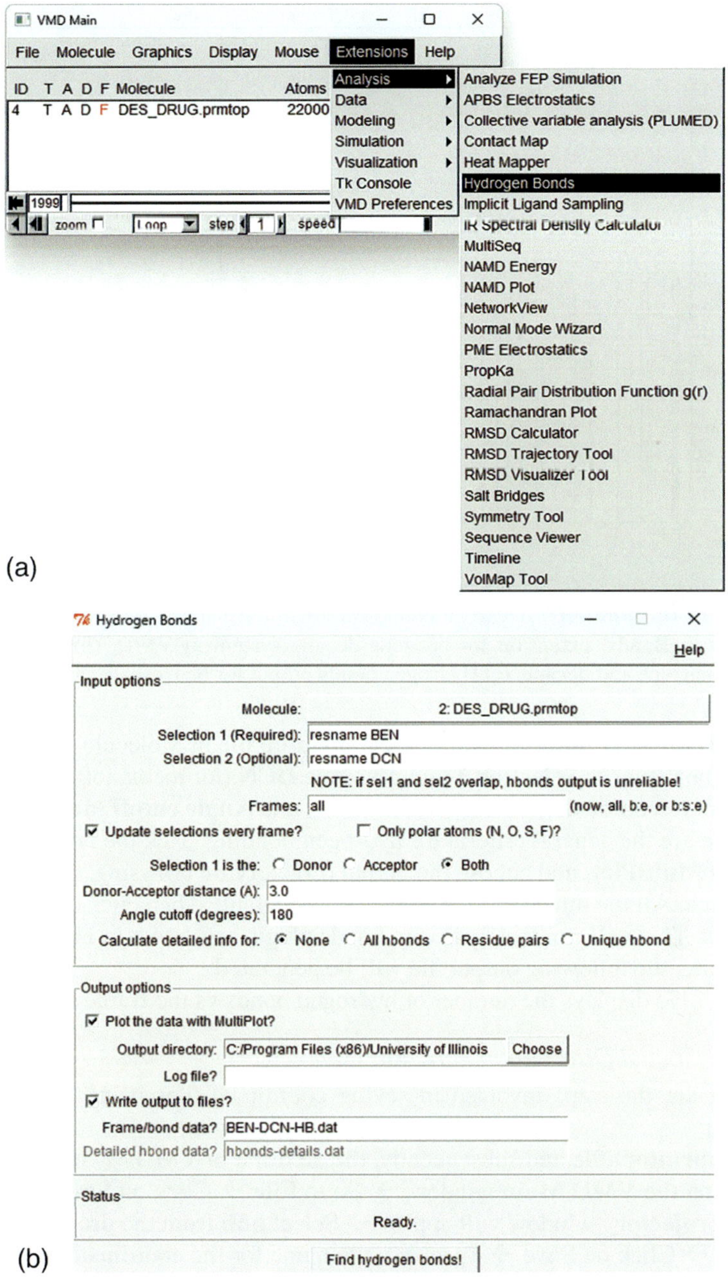

FIGURE 7.17 Hydrogen bonds protocol for benzene-decane system in VMD. (a) 'Hydrogen Bonds' extension for benzene-decane system in VMD. (b) Hydrogen bond inputs for benzene and decane. (c) Hydrogen bonds output for benzene-decane system.

(Continued)

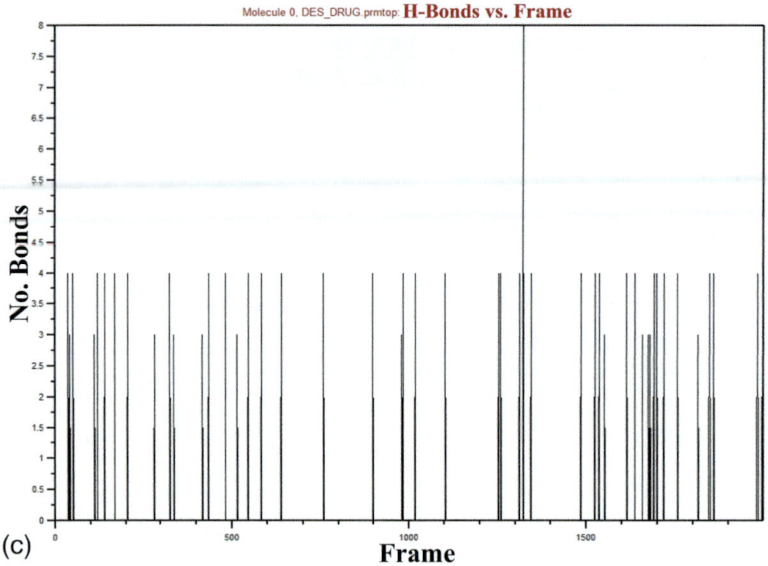

(c)

FIGURE 7.17 (*Continued*) Hydrogen bonds protocol for benzene-decane system in VMD. (a) 'Hydrogen Bonds' extension for benzene-decane system in VMD. (b) Hydrogen bond inputs for benzene and decane. (c) Hydrogen bonds output for benzene-decane system.

In the **Hydrogen Bonds** window, select **.prmtop** file in Molecule, write **resname BEN** (for benzene) as **Selection 1** and **resname DCN** (for methanol) as **Selection 2.** Set **Donor–Acceptor distance** (Å) value as 3.0, and **Angle cutoff (degrees)** value as 180. These are the general criteria for hydrogen bonding. Tick the boxes of **Plot the data with MultiPlot,** and choose the output directory by browsing. Specify the output filename. All the other values will be kept as default. Then click on **Find hydrogen bonds**. This will run the Hydrogen bonds plugin, as shown in Figure 7.17b, and after a while, the following output file will be generated:

Figure 7.17c displays the number of hydrogen bonds vs the frame number plot.

7.2.2.1.4 Spatial Distribution Function (SDF)

Following are the steps involved in saving coordinate files using VMD for SDF generation.

Load **(.prmtop)** file and subsequently, the **.dcd** file in VMD → Select the loaded molecule on the **VMD Main** window → Go to **File** → Click on **Save Coordinates** → **Save Trajectory** window will appear → Select **pdb** from the dropdown menu of **File type** → Click on **Save** → Specify a file name for the coordinate (here we have specified **BDC.pdb**) → Coordinate file (**in pdb format**) will be saved at the specified directory. Figure 7.18a and b displays the respective VMD windows regarding the same.

After saving the coordinate file in pdb format, we continue to do more calculations using the TRAVIS analyzer software in order to get SDF plots. Here are the steps for the calculation of SDF for the acetone–methanol homogeneous system.

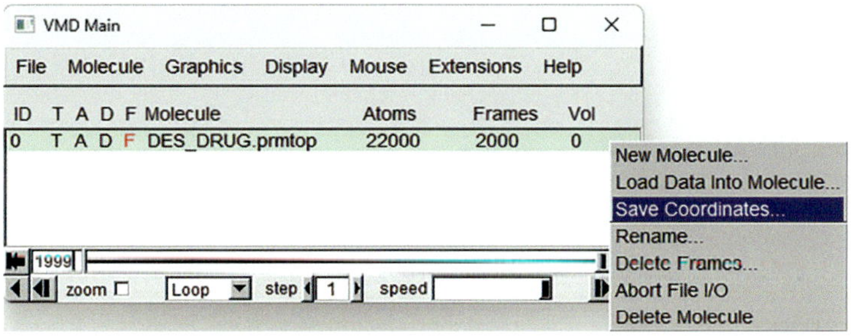

(a)

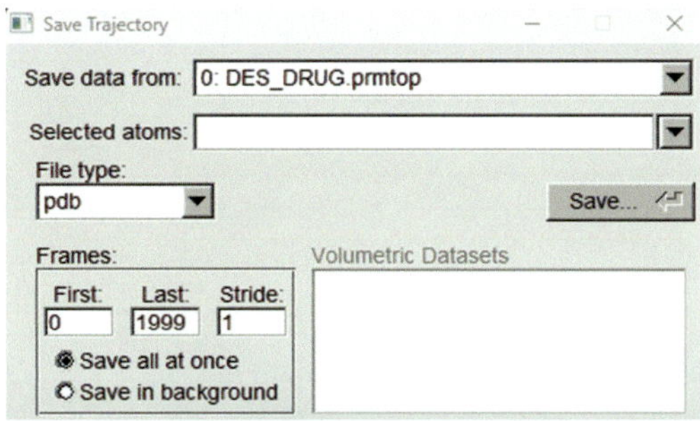

(b)

(c)

FIGURE 7.18 Spatial distribution function (SDF) for benzene-decane system in VMD. (a) 'Save coordinates' extension for benzene-decane system in VMD. (b) Coordinate file saving for benzene-decane system. (c) Trajectory coordinate file loading for benzene-decane system in TRAVIS. (d) Input steps for SDF calculation for benzene-decane system in TRAVIS. (e) Reference molecule of benzene in the VMD display. (f) Isosurface view of reference molecule of benzene in the VMD display. (g) 'Graphical Representations' window for benzene-decane SDF. (h) SDF plot of decane around benzene as central molecule in VMD.

(Continued)

```
! TRAVIS input file
! Created with TRAVIS version compiled at Jul 30 2022 00:22:24
! Source code version: Jul 29 2022
! Input file written at Fri Aug 9 12:49:18 2024.
C:\travis-win64-220729\travis-win64-220729\BDC.pdb
! Use the advanced mode until the analysis selection menu (y/n)? [no]

! Use these values (y) or enter different values (n)? [yes]

! Update cell geometry in every time step (i.e., NPT ensemble) (y) or use fixed cell (n)? [yes]

! Create images of the structural formulas (y/n)? [no]

! Accept these molecules (y) or change something (n)? [yes]

! Which functions to compute (comma separated)?
SDF
! Use the advanced mode for the main part (y/n)? [no]

! Which of the molecules should be the reference molecule (C10H22=1, C6H6=2)?
2
! Please enter three comma-separated reference atoms (e.g. "C1,H2,01"):
C1,H1,H2
! Perform this observation intramolecular (within the reference molecule) (0) or intermolecular (1)? [1]

! Which molecule should be observed (C10H22=1, C6H6=2)?
1
! Which atoms to observe (e.g. "C1,C3-5,H", "*"=all)? [#2]
*
! Please enter maximal observation radius of this SDF (in pm): [1000.0]
1500
! Please enter binning resolution of this SDF per dimension: [100]

! Add a condition to this observation (y/n)? [no]

! Add another observation (y/n)? [no]

! Map a quantity to the surface of the SDF (y/n)? [no]

! Up to which degree should the SDFs be smoothened (0=not at all)? [2]

! Calculate SDF values in nm^-3 (n) or relative to uniform density (y)? [no]

! In which trajectory frame to start processing the trajectory? [1]

! How many trajectory frames to read (from this position on)? [all]

! Use every n-th read trajectory frame for the analysis: [1]
```

(d)

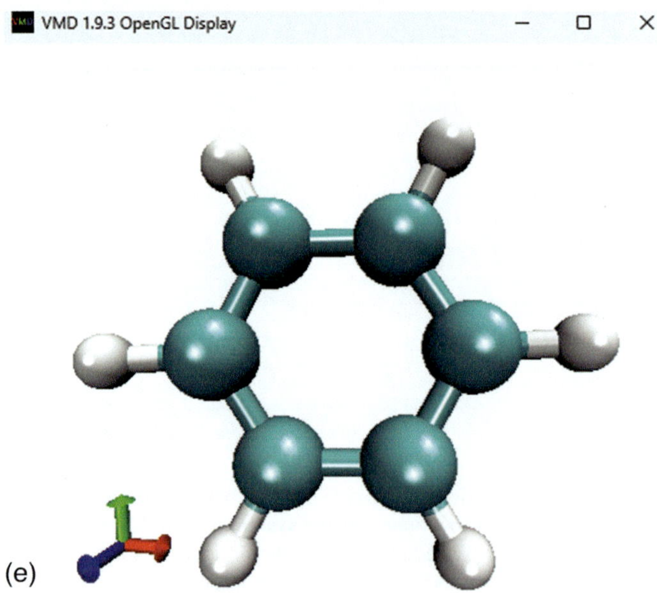

(e)

FIGURE 7.18 (*Continued*) Spatial distribution function (SDF) for benzene-decane system in VMD. (a) 'Save coordinates' extension for benzene-decane system in VMD. (b) Coordinate file saving for benzene-decane system. (c) Trajectory coordinate file loading for benzene-decane system in TRAVIS. (d) Input steps for SDF calculation for benzene-decane system in TRAVIS. (e) Reference molecule of benzene in the VMD display. (f) Isosurface view of reference molecule of benzene in the VMD display. (g) 'Graphical Representations' window for benzene-decane SDF. (h) SDF plot of decane around benzene as central molecule in VMD.

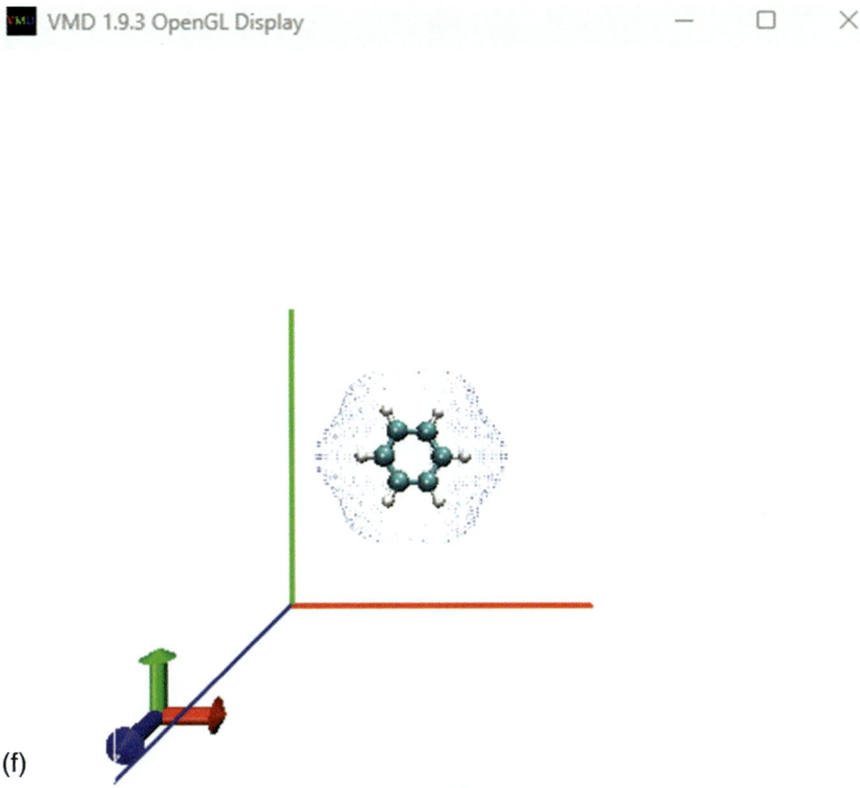

(f)

FIGURE 7.18 (*Continued*) Spatial distribution function (SDF) for benzene-decane system in VMD. (a) 'Save coordinates' extension for benzene-decane system in VMD. (b) Coordinate file saving for benzene-decane system. (c) Trajectory coordinate file loading for benzene-decane system in TRAVIS. (d) Input steps for SDF calculation for benzene-decane system in TRAVIS. (e) Reference molecule of benzene in the VMD display. (f) Isosurface view of reference molecule of benzene in the VMD display. (g) 'Graphical Representations' window for benzene-decane SDF. (h) SDF plot of decane around benzene as central molecule in VMD.

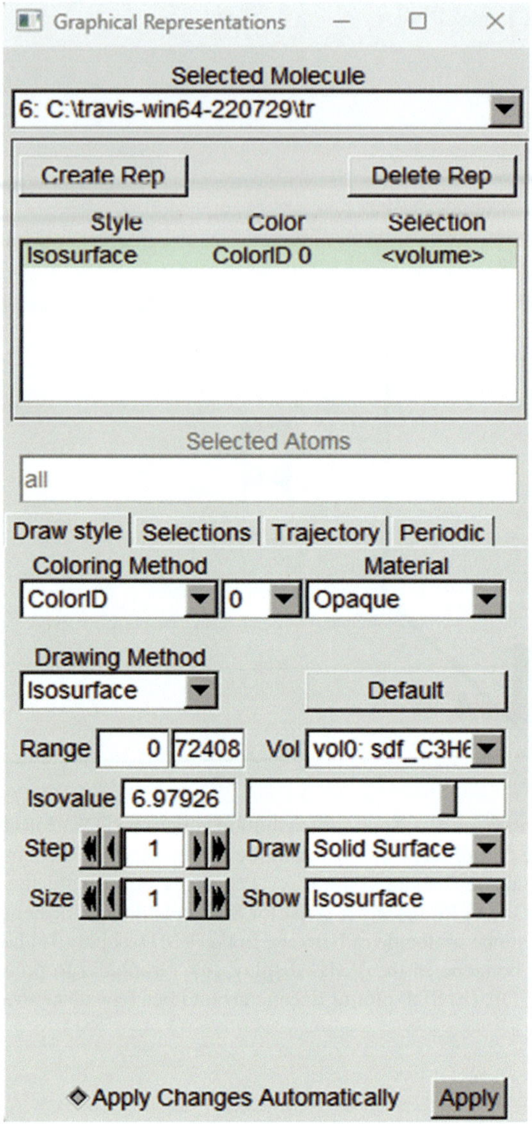

(g)

FIGURE 7.18 (*Continued*) Spatial distribution function (SDF) for benzene-decane system in VMD. (a) 'Save coordinates' extension for benzene-decane system in VMD. (b) Coordinate file saving for benzene-decane system. (c) Trajectory coordinate file loading for benzene-decane system in TRAVIS. (d) Input steps for SDF calculation for benzene-decane system in TRAVIS. (e) Reference molecule of benzene in the VMD display. (f) Isosurface view of reference molecule of benzene in the VMD display. (g) 'Graphical Representations' window for benzene-decane SDF. (h) SDF plot of decane around benzene as central molecule in VMD.

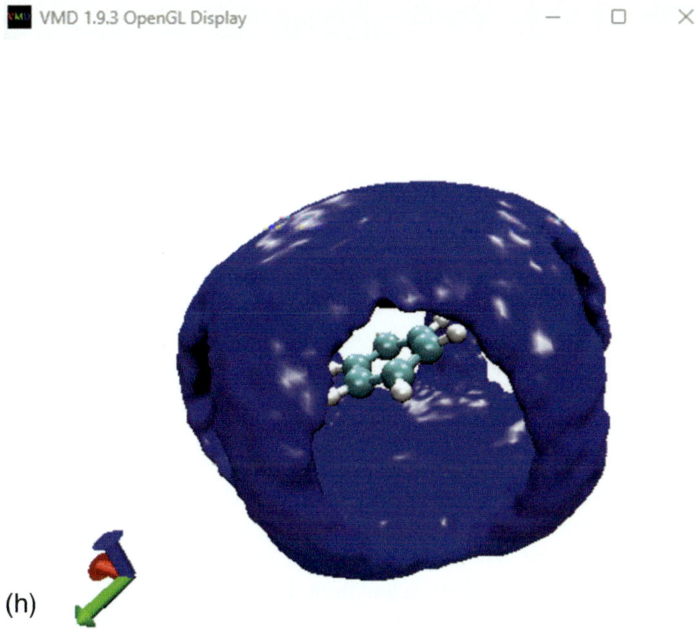

FIGURE 7.18 (*Continued*) Spatial distribution function (SDF) for benzene-decane system in VMD. (a) 'Save coordinates' extension for benzene-decane system in VMD. (b) Coordinate file saving for benzene-decane system. (c) Trajectory coordinate file loading for benzene-decane system in TRAVIS. (d) Input steps for SDF calculation for benzene-decane system in TRAVIS. (e) Reference molecule of benzene in the VMD display. (f) Isosurface view of reference molecule of benzene in the VMD display. (g) 'Graphical Representations' window for benzene-decane SDF. (h) SDF plot of decane around benzene as central molecule in VMD.

Open **travis.exe** → Drag and drop the coordinate file (BDC.pdb) to the travis window → Press **Enter** → BDC.pdb file will be loaded, as referenced by Figure 7.18c.

Figure 7.18d shows the input commands with input steps to be followed in Travis. Acetone has been taken as the reference molecule and methanol, as the observed molecule.

After the completion of these steps, the Travis window will automatically close and a reference coordinate file (**ref_c6h6_c1_h1_h2.xyz**) will be created in the Travis directory. Along with this, the PLT file (.plt) and Cube file (.cube) will be generated. Using these files, we can obtain the SDF plot through VMD. Here are the steps involved in this.

Open VMD → Drag and drop the coordinate file (**ref_c6h6_c1_h1_h2.xyz**) to VMD Main window → Open **Graphical Representations** window from **Graphics** tab → Select **CPK** as the **Drawing Method** → The reference molecule of benzene will appear in the VMD display (Figure 7.18e)

To continue the process, drag and drop the coordinate file Open VMD → Drag and drop the PLT file (.plt) to **VMD Main** window → Initially, it appears as Figure 7.18f → Open the **Graphical Representations** window → Select **ColorID 0** to represent the color "**blue**" for decane isosurface → Adjust the **Isovalue** using the adjustment bar for proper visualization → Select **Solid Surface** and **Isosurface** to specify **Draw** and **Show** options, respectively.

The graphical representations window is presented in Figure 7.18g, and the final SDF plot is displayed in Figure 7.18h.

The blue iso-surfaces surrounding the benzene molecule indicate the probable locations of the decane molecules. These regions suggest areas with a high likelihood of decane molecules. Surfaces generated with higher iso values generally indicate areas of greater density or probability. This suggests that the property being measured, such as the presence of a solvent, is more concentrated or more likely to be found in these regions. The distribution is not uniform, indicating preferential solvation sites around the acetone molecule.

7.2.2.1.5 Root Mean Square Deviation (RMSD)

In the **RMSD Visualizer** window, select **.prmtop** file in Molecule, write **resname BEN** (for benzene) in **Atom Selection** tab. Click on **RMSD.** RMSD result will generate for benzene molecules in the **result** window. Similarly, **write resname DCN** (for decane) in **Atom Selection** tab. Click on **RMSD.** RMSD result will generate for decane molecules in the **result** window. Select both the results and click on **Plot result** tab. The RMSD plots for benzene and decane will be generated, as shown in Figure 7.19b.

(a)
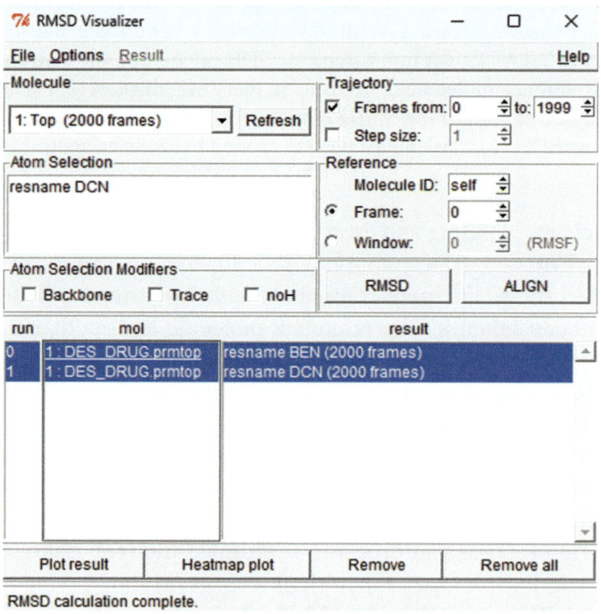

FIGURE 7.19 Root Mean Square Deviation (RMSD) protocol for benzene-decane system in VMD. (a) RMSD visualizer window for benzene-decane system in VMD. (b) RMSD plot for benzene (blue)-decane (red) system.

(Continued)

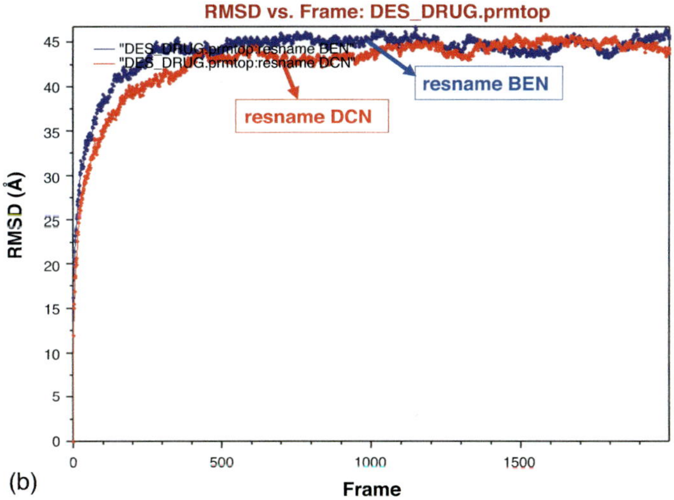

(b)

FIGURE 7.19 (*Continued*) Root Mean Square Deviation (RMSD) protocol for benzene-decane system in VMD. (a) RMSD visualizer window for benzene-decane system in VMD. (b) RMSD plot for benzene (blue)-decane (red) system.

This RMSD plot depicts the variations in molecular structure between two chemical systems, denoted as "BEN" and "DCN", over a molecular dynamics simulation. After around 500 frames, the root mean square deviation (RMSD) values reach a stable state, with the "BEN" molecule settling at around 45 Å and the "DCN" molecule at around 43 Å. The disparity in ultimate RMSD values indicates that the two systems experience varying degrees of conformational alterations; however, both eventually achieve a state of equilibrium where structural fluctuations are minimized. The figure demonstrates that while both systems are dynamic, they eventually settle into rather stable configurations after undergoing early structural adaptations.

7.2.3 (Benzene+Dodecane)

Figure 7.20a presents the chemical structure with atomic notation for benzene and dodecane molecule.

Load the parameter file (.prmtop) and the trajectory file (.dcd) for the benzene–dodecane system in VMD. Adjust the parameters in **Graphical Representations** window setting **resname BEN** and **resname DDC,** for benzene and decane, respectively**.** Select **ColorID** from the dropdown menu of **Coloring Method** tab and **CPK** as the **Drawing Method**. We have assigned **ColorID 1** to represent the color "**red"** for benzene molecules, and **ColorID 0** to represent the color "**blue**" for dodecane molecules, as shown in Figure 7.20c. Other parameters, such as sphere scale, sphere resolution, bond scale, and bond resolution, can be optimized as per the required visual representation for the respective systems.

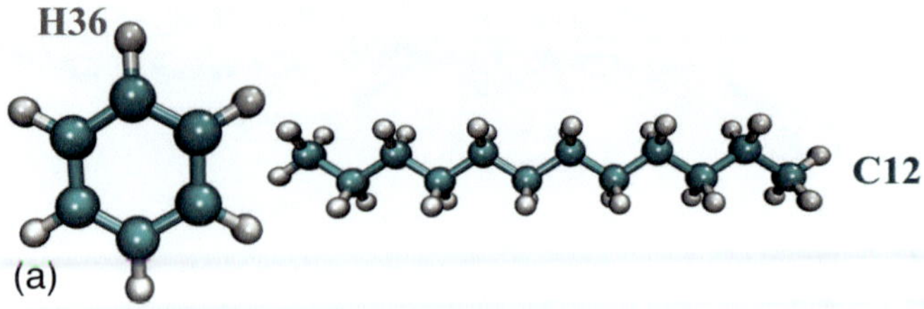

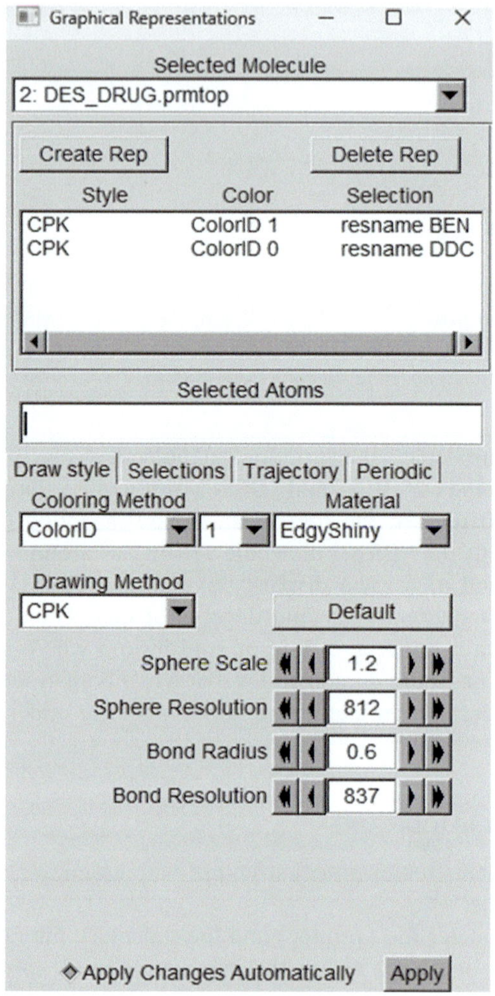

FIGURE 7.20 Configuration of benzene-dodecane system. (a) Chemical structures of benzene and dodecane. (b) 'Graphical Representations' window for benzene-dodecane system. (c) System box visualization (benzene: red, dodecane: blue).

(Continued)

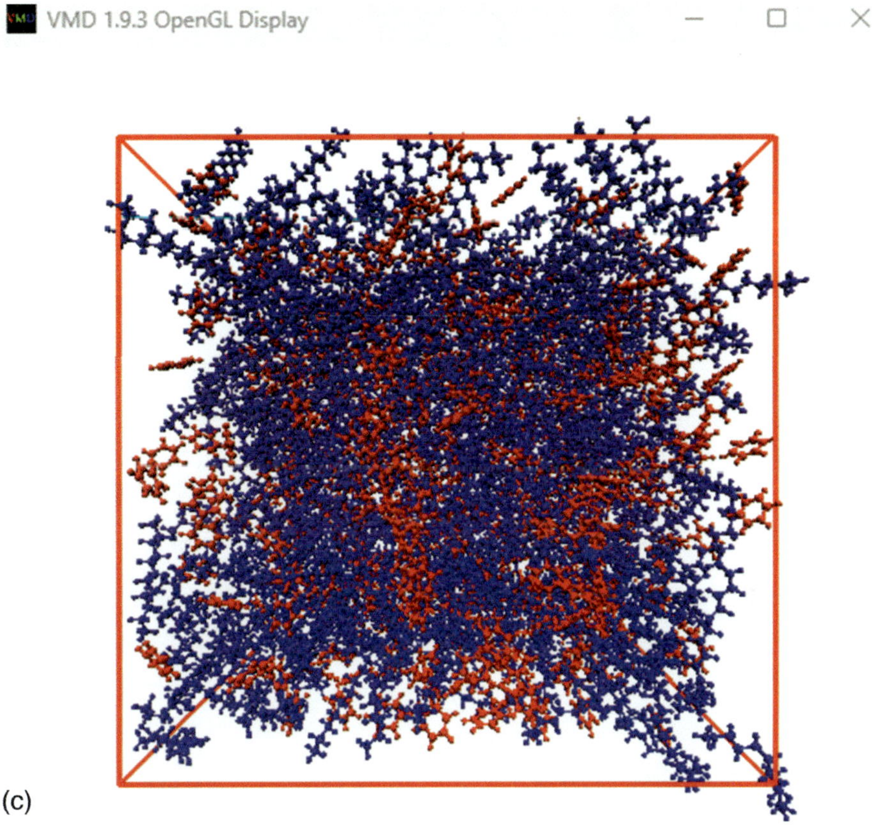

(c)

FIGURE 7.20 (*Continued*) Configuration of benzene-dodecane system. (a) Chemical structures of benzene and dodecane. (b) 'Graphical Representations' window for benzene-dodecane system. (c) System box visualization (benzene: red, dodecane: blue).

7.2.3.1 Structural Properties

7.2.3.1.1 Radial Pair Distribution Function g(r)

Click on **Extensions** → Go to **Analysis** → Select **Radial Pair Distribution Function *g*(*r*)** from the dropdown list → **Radial Pair Distribution Function *g*(*r*)** window will appear, as shown in Figure 7.21b.

In the **Radial Pair Distribution Function *g*(*r*)** window, select **.prmtop** file in Molecule, write **name C12** (for dodecane) as **Selection 1** and **name H36** (for benzene) as **Selection 2.** Set **max. r.** value as 10.0. This value can be set according to the box dimension (in Å). In general, half of the box size is considered. Tick the boxes of **"Use PBC"** and **"Display *g*(*r*)"** . All the other values will be kept as default. Then click on **Compute *g*(*r*)**. This will run the RDF plugin and after a while the following output file will be generated:

As can be observed in Figure 7.21c, dodecane molecules obtained a higher $g(r)$ peak at around 3.3 Å, forming a first solvation shell. The $g(r)$ values peaked to 11.5,

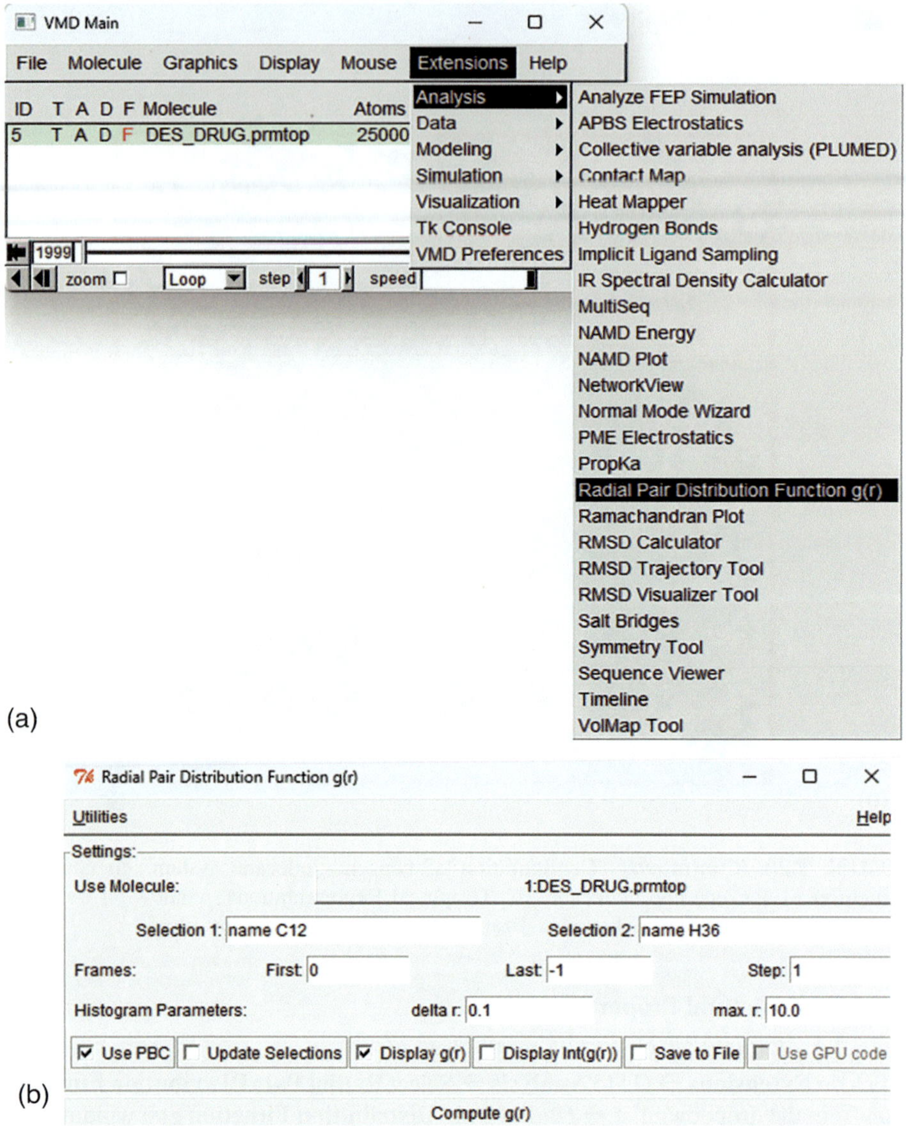

FIGURE 7.21 Radial distribution function (RDF) for benzene-dodecane system in VMD. (a) 'Radial Pair Distribution Function g(r)' extension for benzene-dodecane system. (b) Radial pair distribution function inputs for benzene and dodecane. (c) RDF plot for dodecane (C12)-benzene (H36).

(Continued)

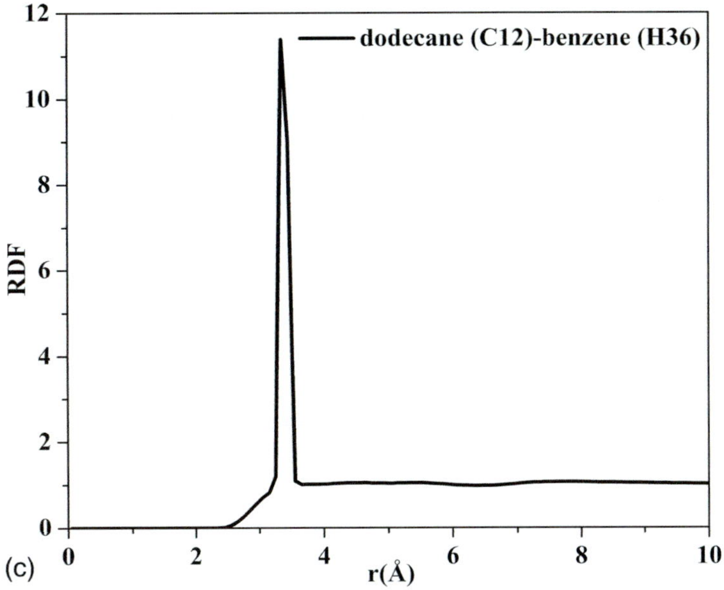

FIGURE 7.21 (*Continued*) Radial distribution function (RDF) for benzene-dodecane system in VMD. (a) 'Radial Pair Distribution Function g(r)' extension for benzene-dodecane system. (b) Radial pair distribution function inputs for benzene and dodecane. (c) RDF plot for dodecane (C12)- benzene (H36).

suggesting a higher coordination number. Both dodecane and benzene are nonpolar entities. The two nonpolar groups interact by dispersion interaction, and thus a higher $g(r)$ peak is observed. No further presence of any solvation shell is witnessed, suggesting the absence of long-range interactions among the benzene and dodecane molecules, a possible result of steric effect.

7.2.3.1.2 Nonbonded Interaction Energy

For nonbonded interaction energy calculation in VMD, the following steps are followed:

Click on **Extensions** → Go to **Analysis** → Select **NAMD Energy** from the drop-down list → The **NAMD Energy** window will appear.

In **NAMD Energy** window, select **.prmtop** file in Molecule, write **resname BEN** (for benzene) as **Selection 1** and **resname DDC** (for dodecane) as **Selection 2.** Tick the boxes of **VDW, Elec, and Nonbond** that respectively indicate van der Waals, electrostatic, and total nonbonded interactions. Set **Output File** name (e.g. BEN-DDC-IE). Select the parameter file and XSC file by browsing. All the other values will be kept as default. Then click on **Run NAMD Energy.** This will run the NAMD energy plugin, and after a while, the following output (Figure 7.22b) file will be generated:

For the benzene–dodecane system, the electrostatic and the vdW components of interaction energy are found to be −0.01 and −4.14 kcal/mol, respectively. The total nonbonded interaction energy is −4.15 kcal/mol.

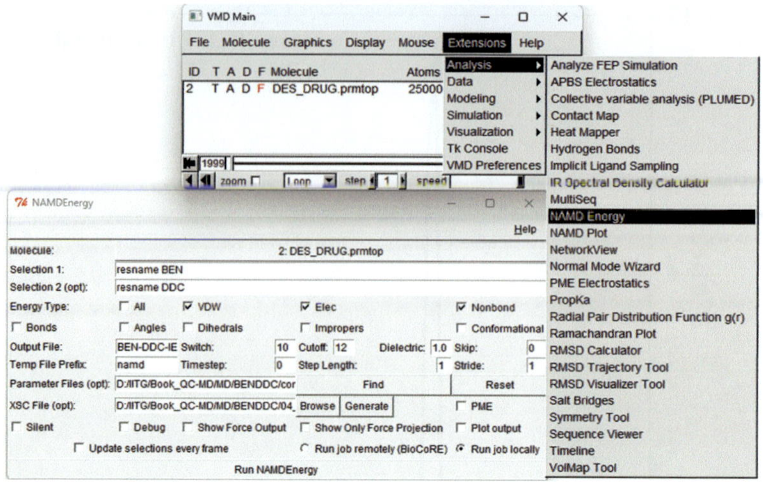

(a)

BEN-DDC-IE

Frame	Time	Elec	VdW	Nonbond	Total
0	0	-3.8561	-4160.81	-4164.67	-4164.67
1	1	-3.6032	-4100.07	-4103.67	-4103.67
2	2	-8.3727	-4156.86	-4165.23	-4165.23
3	3	-6.7177	-4097.56	-4104.27	-4104.27
4	4	-2.4811	-4143.55	-4146.03	-4146.03
5	5	-0.316	-4136.76	-4137.08	-4137.08
6	6	-3.3221	-4181.44	-4184.76	-4184.76
7	7	-4.2031	-4132.34	-4136.55	-4136.55
8	8	-8.2306	-4136.96	-4145.19	-4145.19
9	9	-8.9449	-4119.24	-4128.18	-4128.18
10	10	-4.1747	-4133.68	-4137.85	-4137.85
11	11	-8.2029	-4122.12	-4130.32	-4130.32
12	12	-7.389	-4123.08	-4130.47	-4130.47
13	13	-6.5643	-4173.45	-4180.01	-4180.01
14	14	-4.3773	-4164.52	-4168.89	-4168.89
15	15	-4.5659	-4093.83	-4098.39	-4098.39
16	16	-3.1637	-4106.48	-4109.64	-4109.64
17	17	-10.4366	-4202.73	-4213.16	-4213.16
18	18	-6.9286	-4156.87	-4163.8	-4163.8
19	19	-6.7137	-4132.59	-4139.3	-4139.3
20	20	-4.5386	-4161.83	-4166.37	-4166.37
21	21	-7.9128	-4143.53	-4151.44	-4151.44
22	22	-11.2816	-4139.18	-4150.47	-4150.47
23	23	-0.0599	-4104.22	-4104.28	-4104.28
24	24	-7.246	-4168.31	-4175.55	-4175.55
25	25	-8.6102	-4162.84	-4171.45	-4171.45
26	26	-8.0267	-4157.49	-4165.52	-4165.52
27	27	-4.8799	-4156.4	-4161.28	-4161.28
28	28	-6.5873	-4138.24	-4144.83	-4144.83
29	29	-3.6011	-4161.8	-4165.4	-4165.4
30	30	-8.8323	-4150.77	-4159.6	-4159.6

(b)

FIGURE 7.22 Non-bonded interaction energy (NAMD energy) protocol for benzene-dodecane system in VMD. (a) NAMD energy inputs for benzene-dodecane system. (b) NAMD energy output for benzene-dodecane system.

7.2.3.1.3 Hydrogen Bonding

Click on **Extensions** → Go to **Analysis** → Select **Hydrogen Bonds** from the drop-down list → **Hydrogen Bonds** window will appear, as shown in Figure 7.23a.

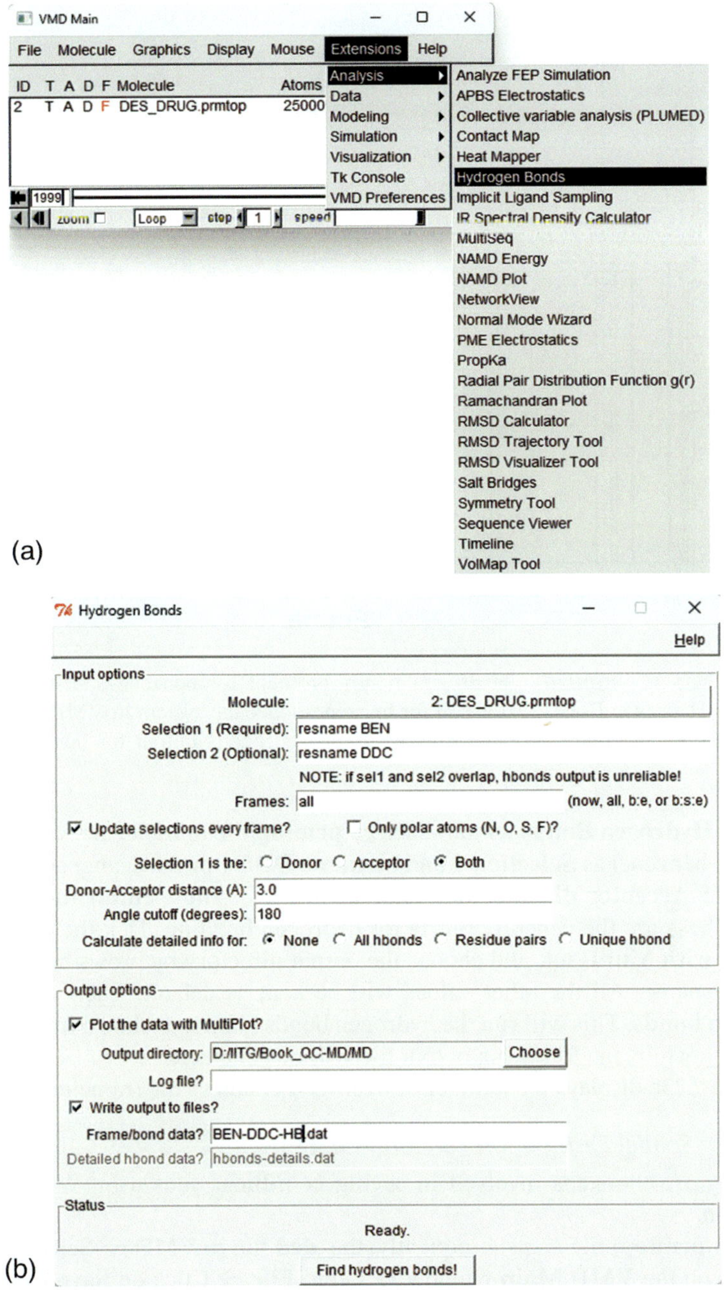

FIGURE 7.23 Hydrogen bonds protocol for benzene-dodecane system in VMD. (a) 'Hydrogen Bonds' extension for benzene-dodecane system in VMD. (b) Hydrogen bond inputs for benzene and dodecane. (c) Hydrogen bonds output for benzene-dodecane system.

(Continued)

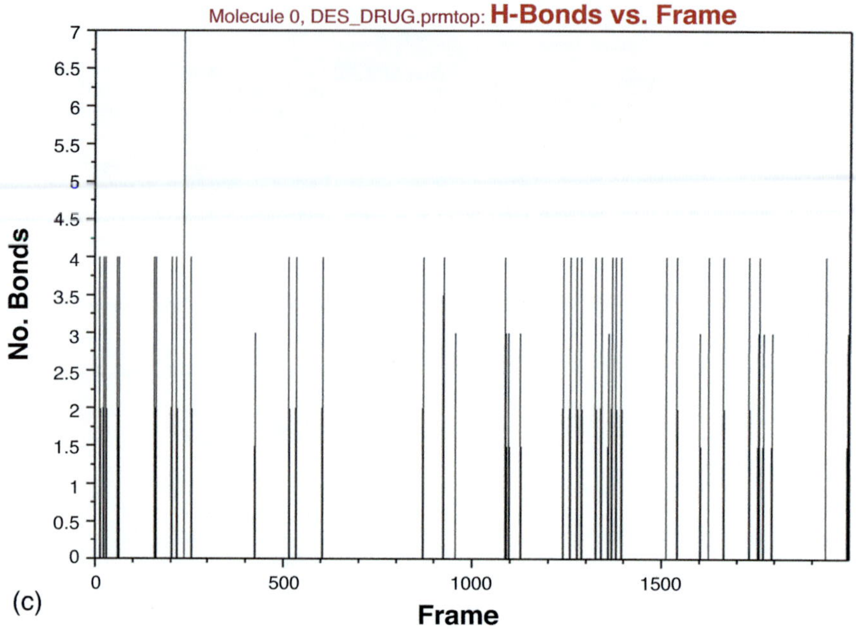

(c)

FIGURE 7.23 (*Continued*) Hydrogen bonds protocol for benzene-dodecane system in VMD. (a) 'Hydrogen Bonds' extension for benzene-dodecane system in VMD. (b) Hydrogen bond inputs for benzene and dodecane. (c) Hydrogen bonds output for benzene-dodecane system.

In the **Hydrogen Bonds** window, select **.prmtop** file in Molecule, write **resname BEN** (for benzene) as **Selection 1** and **resname DDC** (for dodecane) as **Selection 2.** Set **Donor-Acceptor distance** (Å) value as 3.0, and **Angle cutoff (degrees)** value as 180. These are the general criteria for hydrogen bonding. Tick the boxes of **Plot the data with MultiPlot,** and choose the output directory by browsing. Specify the output filename. All the other values will be kept as default. Then click on **Find hydrogen bonds**. This will run the hydrogen bonds plugin, as shown in Figure 7.23b, and after a while, the following output file will be generated:

Figure 7.23c displays the number of hydrogen bonds vs the frame number plot.

7.2.3.1.4 Spatial Distribution Function (SDF)

Following are the steps involved in saving coordinate files using VMD for SDF generation.

Load **(.prmtop)** file and subsequently, the **.dcd** file in VMD → Select the loaded molecule on the **VMD Main** window → Go to **File** → Click on **Save Coordinates** → **Save Trajectory** window will appear → Select **pdb** from the dropdown menu of **File type** → Click on **Save** → Specify a file name for the coordinate (here we have specified **BDD.pdb**) → Coordinate file **(in pdb format)** will be saved at the specified directory. Figure 7.24a and b displays the respective VMD windows regarding the same.

After saving the coordinate file in pdb format, we continue to do more calculations using the TRAVIS analyzer software in order to get SDF plots. Here are the steps for the calculation of SDF for the acetone–methanol homogeneous system.

Open **travis.exe** → Drag and drop the coordinate file (BDD.pdb) to the travis window → Press **Enter** → BDD.pdb file will be loaded as referenced by Figure 7.24c.

Figure 7.24d shows the input commands with input steps to be followed in Travis. Acetone has been taken as the reference molecule and methanol, as the observed molecule.

After the completion of these steps, the Travis window will automatically close, and a reference coordinate file (**ref_c6h6_c1_h1_h2.xyz**) will be created in the Travis directory. Along with this, the PLT file (.plt) and Cube file (.cube) will be generated. Using these files, we can obtain the SDF plot through VMD. Here are the steps involved in this.

Open VMD → Drag and drop the coordinate file (**ref_c6h6_c1_h1_h2.xyz**) to VMD Main window → Open **Graphical Representations** window from **Graphics**

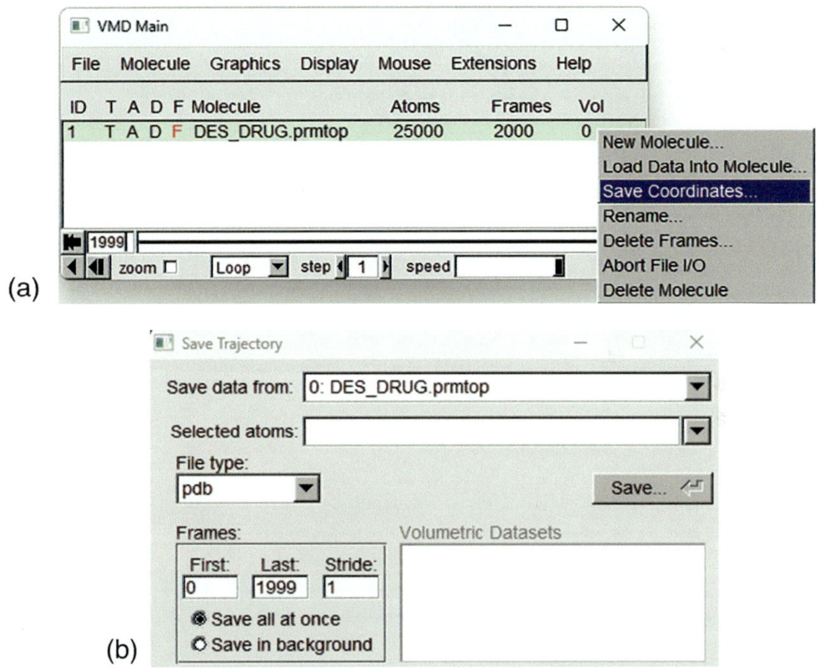

FIGURE 7.24 Spatial distribution function (SDF) for benzene-dodecane system in VMD. (a) 'Save coordinates' extension for benzene-dodecane system in VMD. (b) Coordinate file saving for benzene-dodecane system. (c) Trajectory coordinate file loading for benzene-dodecane system in TRAVIS. (d) Input steps for SDF calculation for benzene-dodecane system in TRAVIS. (e) Reference molecule of benzene in the VMD display. (f) Isosurface view of reference molecule of benzene in the VMD display. (g) 'Graphical Representations' window for benzene-dodecane SDF. (h) SDF plot of dodecane around benzene as central molecule in VMD.

(Continued)

(c) 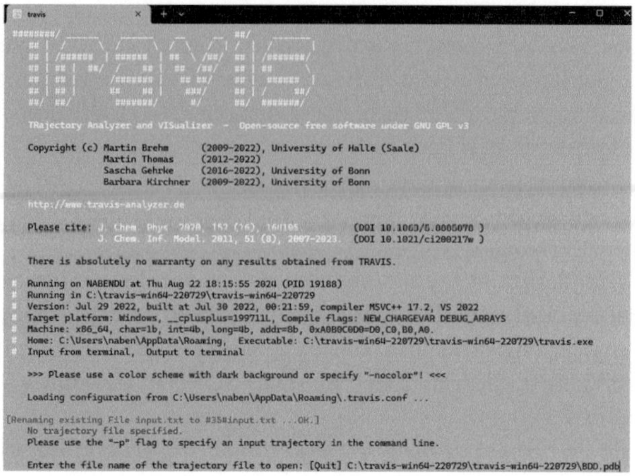

(d)

```
! TRAVIS input file
! Created with TRAVIS version compiled at Jul 30 2022 00:22:24
! Source code version: Jul 29 2022
! Input file written at Sat Aug 10 20:59:47 2024.
C:\travis-win64-220729\travis-win64-220729\BDD.pdb
! Use the advanced mode until the analysis selection menu (y/n)? [no]

! Use these values (y) or enter different values (n)? [yes]

! Update cell geometry in every time step (i.e., NPT ensemble) (y) or use fixed cell (n)? [yes]

! Create images of the structural formulas (y/n)? [no]

! Accept these molecules (y) or change something (n)? [yes]

! Which functions to compute (comma separated)?
SDF
! Use the advanced mode for the main part (y/n)? [no]

! Which of the molecules should be the reference molecule (C12H26=1, C6H6=2)?
2
! Please emter three comma-separated reference atoms (e.g. "C1,H2,O1"):
C1,H1,H2
! Perform this observation intramolecular (within the reference molecule) (0) or intermolecular (1)? [1]

! Which molecule should be observe (C12H26=1, C6H6=2)?
1
! Which atoms to observe (e.g. "C1,C3-5,H", "*"=all)? [#2]
*
! Please enter maximal observation radius of this SDF (in pm): [1000.0] |

! Please enter binning resolution of this SDF per dimension: [100]

! Add a condition to this observation (y/n)? [no]

! Add another observation (y/n)? [no]

! Map a quantity to the surface of the SDF (y/n)? [no]

! Up to which degree should the SDFs be smoothened (0=not at all)? [2]

! Calculate SDF values in nm^-3 (n) or relative to uniform density (y)? [no]

! In which trajectory frame to start processing the trajectory? [1]

! How many trajectory frames to read (from this position on)? [all]

! Use every n-th read trajectory frame for the analysis: [1]
```

FIGURE 7.24 (*Continued*) Spatial distribution function (SDF) for benzene-dodecane system in VMD. (a) 'Save coordinates' extension for benzene-dodecane system in VMD. (b) Coordinate file saving for benzene-dodecane system. (c) Trajectory coordinate file loading for benzene-dodecane system in TRAVIS. (d) Input steps for SDF calculation for benzene-dodecane system in TRAVIS. (e) Reference molecule of benzene in the VMD display. (f) Isosurface view of reference molecule of benzene in the VMD display. (g) 'Graphical Representations' window for benzene-dodecane SDF. (h) SDF plot of dodecane around benzene as central molecule in VMD.

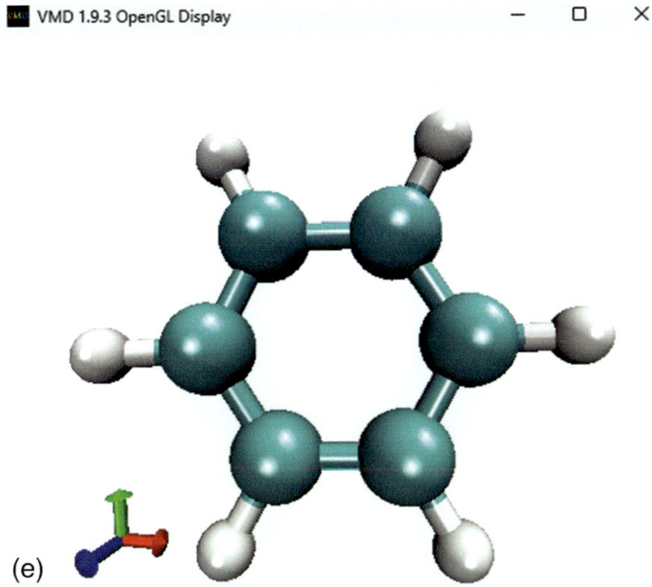

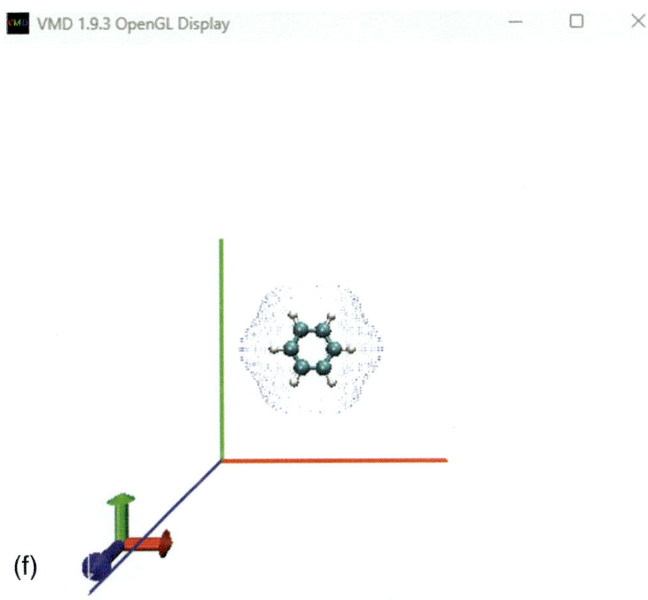

FIGURE 7.24 (*Continued*) Spatial distribution function (SDF) for benzene-dodecane system in VMD. (a) 'Save coordinates' extension for benzene-dodecane system in VMD. (b) Coordinate file saving for benzene-dodecane system. (c) Trajectory coordinate file loading for benzene-dodecane system in TRAVIS. (d) Input steps for SDF calculation for benzene-dodecane system in TRAVIS. (e) Reference molecule of benzene in the VMD display. (f) Isosurface view of reference molecule of benzene in the VMD display. (g) 'Graphical Representations' window for benzene-dodecane SDF. (h) SDF plot of dodecane around benzene as central molecule in VMD.

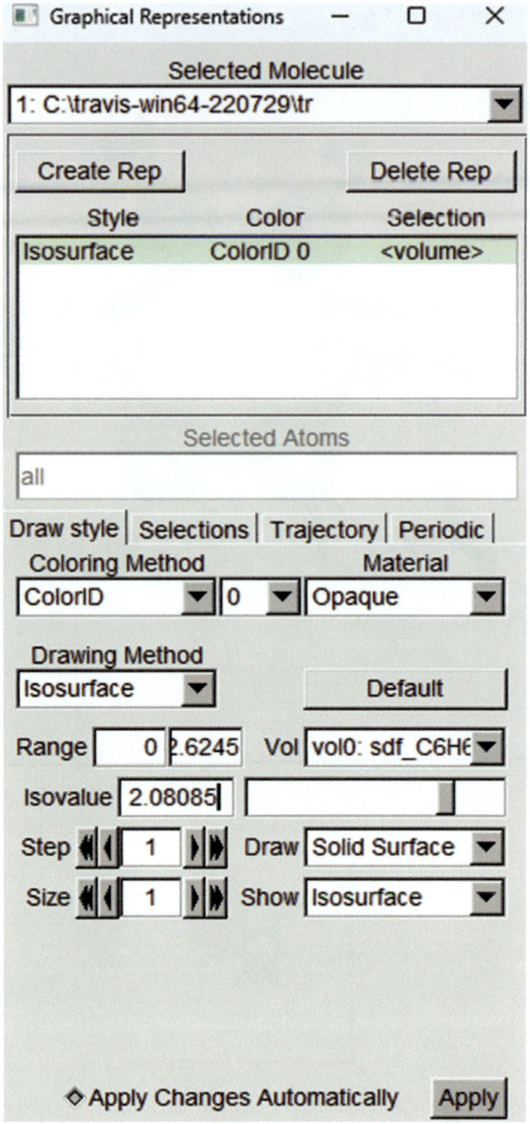

(g)

FIGURE 7.24 (*Continued*) Spatial distribution function (SDF) for benzene-dodecane system in VMD. (a) 'Save coordinates' extension for benzene-dodecane system in VMD. (b) Coordinate file saving for benzene-dodecane system. (c) Trajectory coordinate file loading for benzene-dodecane system in TRAVIS. (d) Input steps for SDF calculation for benzene-dodecane system in TRAVIS. (e) Reference molecule of benzene in the VMD display. (f) Isosurface view of reference molecule of benzene in the VMD display. (g) 'Graphical Representations' window for benzene-dodecane SDF. (h) SDF plot of dodecane around benzene as central molecule in VMD.

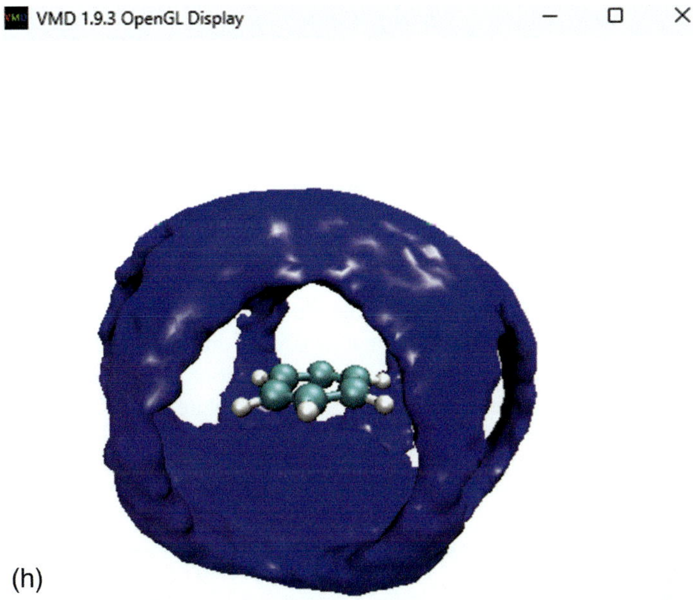

FIGURE 7.24 (*Continued*) Spatial distribution function (SDF) for benzene-dodecane system in VMD. (a) 'Save coordinates' extension for benzene-dodecane system in VMD. (b) Coordinate file saving for benzene-dodecane system. (c) Trajectory coordinate file loading for benzene-dodecane system in TRAVIS. (d) Input steps for SDF calculation for benzene-dodecane system in TRAVIS. (e) Reference molecule of benzene in the VMD display. (f) Isosurface view of reference molecule of benzene in the VMD display. (g) 'Graphical Representations' window for benzene-dodecane SDF. (h) SDF plot of dodecane around benzene as central molecule in VMD.

tab → Select **CPK** as the **Drawing Method** → The reference molecule of benzene will appear in the VMD display (Figure 7.24e).

To continue the process, drag and drop the coordinate file Open VMD → Drag and drop the PLT file (.plt) to **VMD Main** window → Initially, it appears as Figure 7.24f → Open the **Graphical Representations** window → Select **ColorID 0** to represent the color "**blue**" for dodecane isosurface → Adjust the **Isovalue** using the adjustment bar for proper visualization → Select **Solid Surface** and **Isosurface** to specify **Draw** and **Show** options, respectively.

The graphical representations window is presented in Figure 7.24g, and the final SDF plot is displayed in Figure 7.24h.

The blue iso-surfaces surrounding the benzene molecule indicate the probable locations of the dodecane molecules. These regions suggest areas with a high likelihood of dodecane molecules. Surfaces generated with higher iso values generally indicate areas of greater density or probability. This suggests that the property being measured, such as the presence of a solvent, is more concentrated or more likely to be found in these regions. The distribution is not uniform, indicating preferential solvation sites around the acetone molecule.

7.2.3.1.5 Root Mean Square Deviation (RMSD)

In the **RMSD Visualizer** window, select **.prmtop** file in Molecule, write **resname BEN** (for benzene) in **Atom Selection** tab. Click on **RMSD.** RMSD result will generate for benzene molecules in the **result** window. Similarly, **write resname DDC** (for dodecane) in **Atom Selection** tab**.** Click on **RMSD.** RMSD result will generate for decane molecules in the **result** window. Select both the results and click on **Plot result** tab. The RMSD plots for benzene and decane will be generated, as shown in Figure 7.25b.

The RMSD plot compares the structural deviations of two molecular systems, "BEN" (blue) and "DDC" (red), over time during a molecular dynamics simulation. Both systems exhibit a rapid increase in RMSD initially, indicating significant structural adjustments from their starting configurations. After approximately 500 frames, both systems stabilize, with the RMSD values plateauing. The "BEN" system

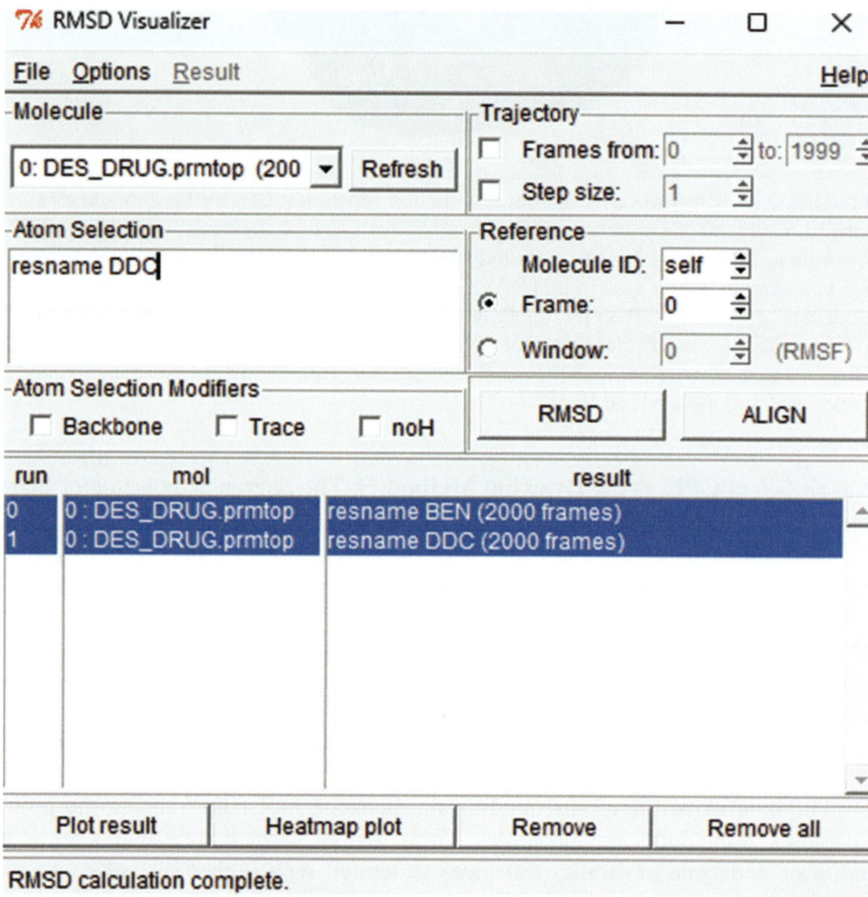

(a)

FIGURE 7.25 Root Mean Square Deviation (RMSD) protocol for benzene-dodecane system in VMD. (a) RMSD visualizer window for benzene-dodecane system in VMD. (b) RMSD plot for benzene (blue)-dodecane (red) system.

(Continued)

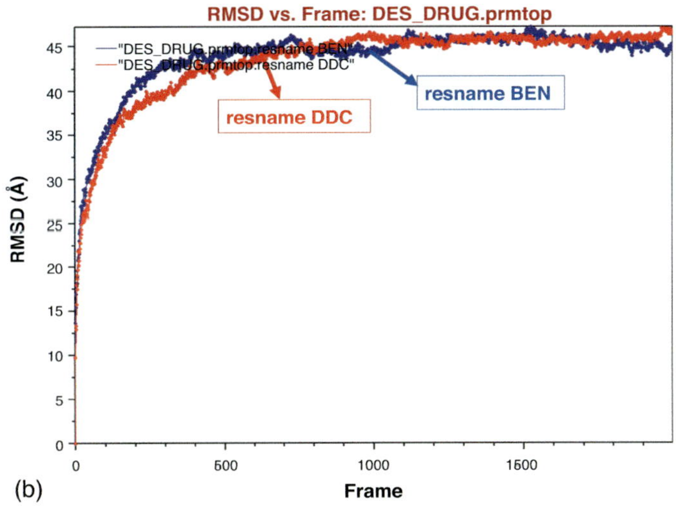

FIGURE 7.25 (*Continued*) Root Mean Square Deviation (RMSD) protocol for benzene-dodecane system in VMD. (a) RMSD visualizer window for benzene-dodecane system in VMD. (b) RMSD plot for benzene (blue)-dodecane (red) system.

stabilizes around 45 Å, while the "DDC" system stabilizes slightly lower, around 43 Å, suggesting that "BEN" undergoes marginally greater structural deviation. This behavior indicates that both systems reach a stable conformation, with "BEN" showing slightly more variation from its reference structure compared to "DDC".

7.3 HETEROGENEOUS SYSTEM (BENZENE–WATER)

The current section deals with the post-processing and analysis of the heterogeneous system of benzene and water. Structural properties, interaction energy analysis, hydrogen bonding, and diffusive properties will be discussed in this part.

Load the parameter file (.prmtop) and the trajectory file (.dcd) for the benzene-water system in VMD. Adjust the parameters in **Graphical Representations** window, setting **resname BEN** and **resname wtr.** Select **ColorID** from the dropdown menu of the **Coloring Method** tab and **CPK** as the **Drawing Method**. We have assigned **ColorID 1** to represent the color "**red**" for benzene molecules, and **ColorID 0** to represent the color "**blue"** for water molecules, as shown in Figure 7.26c. Other parameters, such as sphere scale, sphere resolution, bond scale, and bond resolution, can be optimized as per the required visual representation for the respective systems.

7.3.1 Structural Properties

Structural properties in NAMD are the characteristics of molecular systems that can be analyzed through molecular dynamics simulations. These properties facilitate understanding the configuration and behavior of molecules over time. In this section, we will focus on two specific structural properties: RDF and SDF.

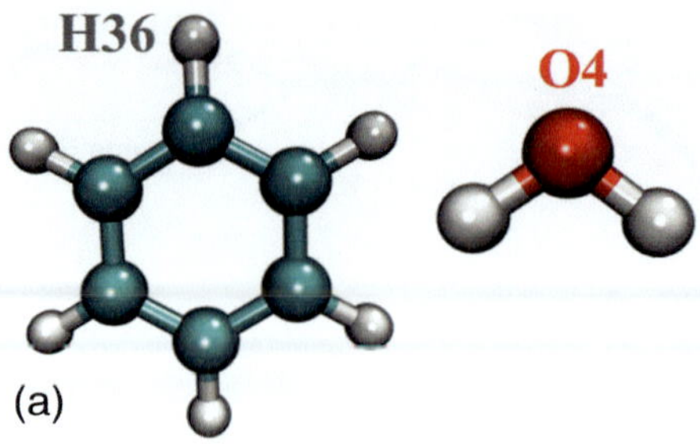

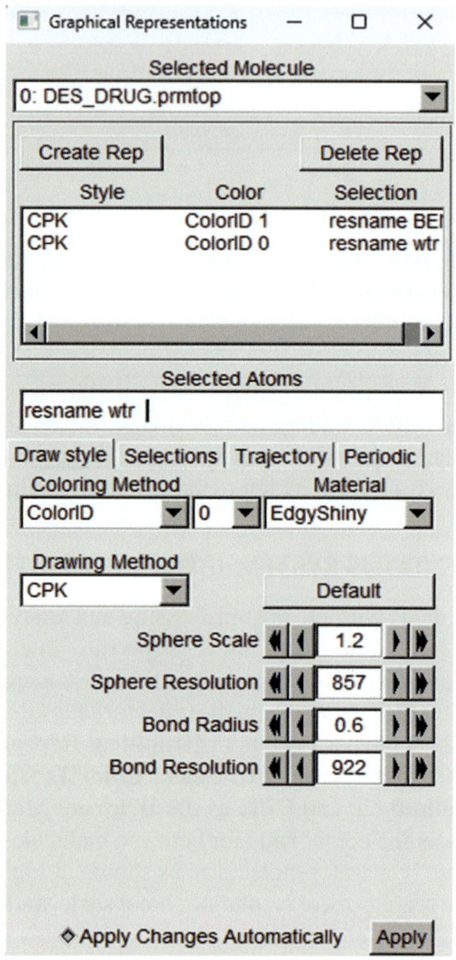

(b)

FIGURE 7.26 Configuration of benzene-water system. (a) Chemical structures of benzene and water. (b) 'Graphical Representations' window for benzene-water system. (c) System box visualization (benzene: red, water: blue).

(Continued)

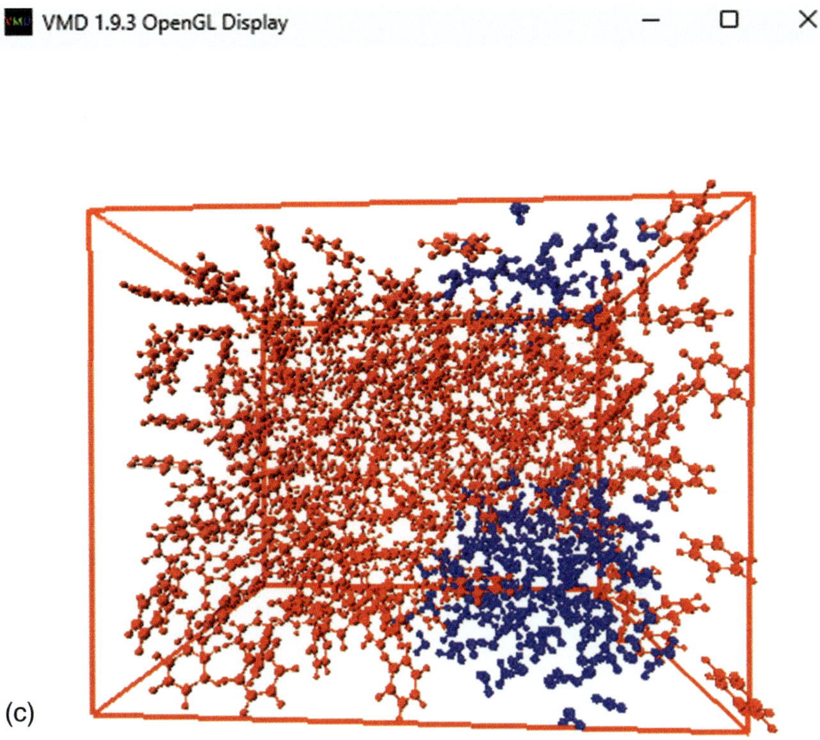

(c)

FIGURE 7.26 (*Continued*) Configuration of benzene-water system. (a) Chemical structures of benzene and water. (b) 'Graphical Representations' window for benzene-water system. (c) System box visualization (benzene: red, water: blue).

7.3.1.1 Radial Pair Distribution Function *g*(*r*)

To obtain the RDF plot in VMD, open the **VMD Main** window → Go to the **Extensions** tab → Select **Analysis** → Click on **Radial Pair Distribution Function *g*(*r*)** from the dropdown list. The steps are displayed in Figure 7.27a. A **Radial Pair Distribution Function *g*(*r*)** window will appear, as shown in Figure 7.27b.

In the **Radial Pair Distribution Function *g*(*r*)** window, select the parameter file (**.prmtop**) in the **Use Molecule** option. In the **Selection 1** and **Selection 2** tab, write "**name O4**" and "**name H36**", respectively. **Max. r.** is set at 15, half of the box size in Å. All the other parameters are kept as default. Click on **Compute *g*(*r*).** Figure 7.27c will appear as the RDF plot.

Figure 7.27c shows that there is no prominent *g*(*r*) peak between water and benzene molecules. Benzene lacks polarity, while water molecules possess a polar nature. These two substances form a heterogeneous system because of their immiscibility, resulting in a system with multiple phases. It is evident that the interaction between benzene and water molecules is minimal, resulting in a low density of water around the benzene molecule. From Figure 7.27c, it is evident that there is no clear correlation between the two components, indicating a lower level of chemical interaction.

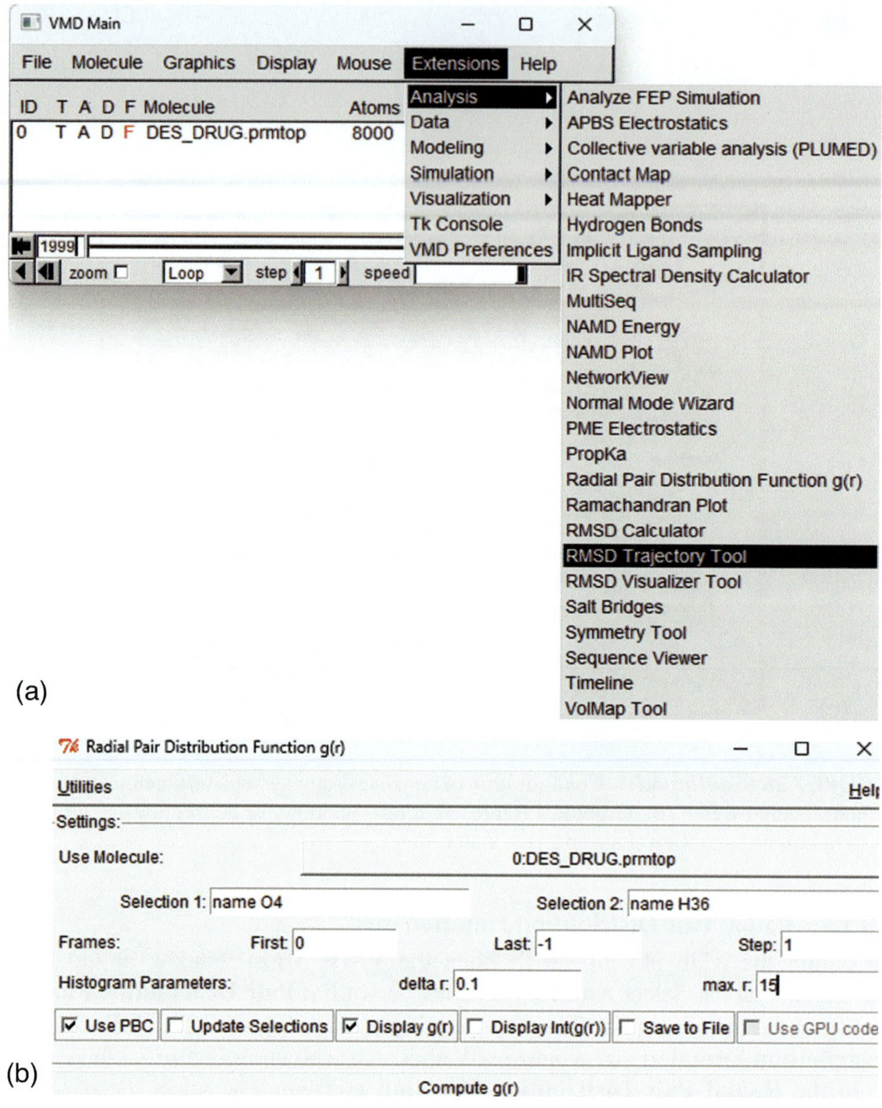

FIGURE 7.27 Radial distribution function (RDF) for benzene-water system in VMD. (a) 'Radial Pair Distribution Function g(r)' extension for benzene-water system. (b) Radial pair distribution function inputs for benzene and water. (c) RDF plot for water (O4)- benzene (H36).

(Continued)

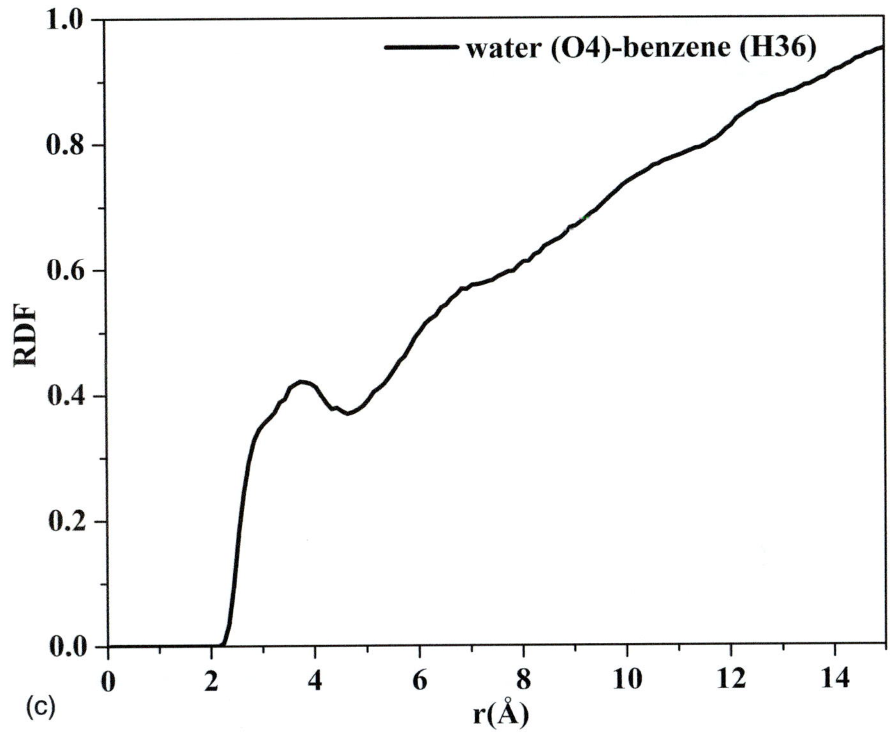

FIGURE 7.27 *(Continued)* Radial distribution function (RDF) for benzene-water system in VMD. (a) 'Radial Pair Distribution Function g(r)' extension for benzene-water system. (b) Radial pair distribution function inputs for benzene and water. (c) RDF plot for water (O4)-benzene (H36).

7.3.1.2 Spatial Distribution Function (SDF)

Following are the steps involved in saving coordinate files using VMD for SDF generation.

Load **(.prmtop)** file and subsequently, the **.dcd** file in VMD → Select the loaded molecule on the **VMD Main** window → Go to **File** → Click on **Save Coordinates** → **Save Trajectory** window will appear → Select **pdb** from the dropdown menu of **File type** → Click on **Save** → Specify a file name for the coordinate (here we have specified **BNW.pdb**) → Coordinate file (**in pdb format**) will be saved at the specified directory. Figure 7.28a and b display the respective VMD windows regarding the same.

After saving the coordinate file in pdb format, we continue to do more calculations using the TRAVIS analyzer software in order to get SDF plots. Here are the steps for the calculation of SDF for the acetone–methanol homogeneous system.

Open **travis.exe** → Drag and drop the coordinate file (BNW.pdb) to the travis window → Press **Enter** → BNW.pdb file will be loaded as referenced by Figure 7.28c.

Figure 7.28d shows the input commands with input steps to be followed in Travis. Benzene has been taken as the reference molecule and water as the observed molecule.

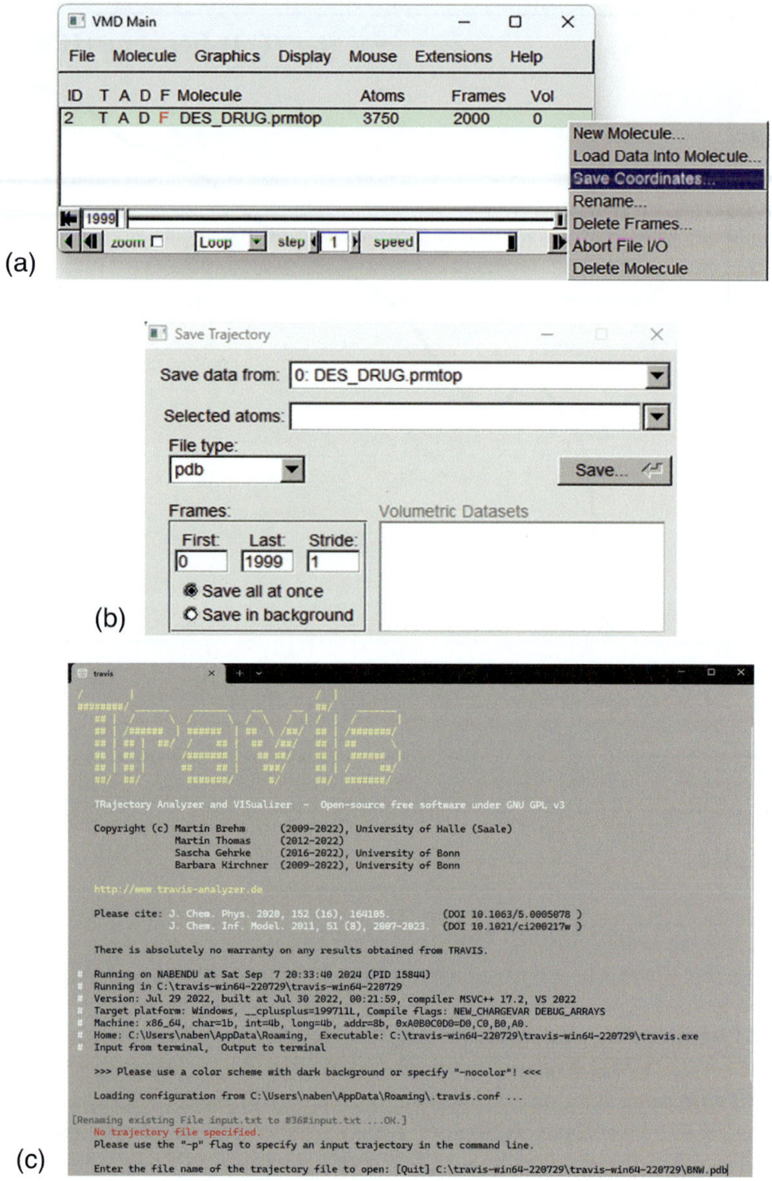

FIGURE 7.28 Spatial distribution function (SDF) for benzene-water system in VMD. (a) 'Save coordinates' extension for benzene-water system in VMD. (b) Coordinate file saving for benzene-water system. (c) Trajectory coordinate file loading for benzene-water system in TRAVIS. (d) Input steps for SDF calculation for benzene-water system in TRAVIS. (e) Reference molecule of benzene in the VMD display. (f) Isosurface view of reference molecule of benzene in the VMD display. (g) 'Graphical Representations' window for benzene-water SDF. (h) SDF plot of water around benzene as central molecule in VMD.

(Continued)

```
C:\travis-win64-220729\travis-win64-220729\BNW.pdb
! Use the advanced mode until the analysis selection menu (y/n)? [no]

! Use these values (y) or enter different values (n)? [yes]

! Update cell geometry in every time step (i.e., NPT ensemble) (y) or use fixed cell (n)? [yes]

! Create images of the structural formulas (y/n)? [no]

! Accept these molecules (y) or change something (n)? [yes]

! Which functions to compute (comma separated)?
sdf
! Use the advanced mode for the main part (y/n)? [no]

! Which of the molecules should be the reference molecule (C6H6=1, H2O=2)?
1
! Please enter three comma-separated reference atoms (e.g. "C1,H2,O1"):
C1,H1,H2
! Perform this observation intramolecular (within the reference molecule) (0) or intermolecular (1)? [1]

! Which molecule should be observed (C6H6=1, H2O=2)?
2
! Which atoms to observe (e.g. "C1,C3-5,H", "*"=all)? [#2]
*
! Please enter maximal observation radius of this SDF (in pm): [1000.0]

! Please enter binning resolution of this SDF per dimension: [100]

! Add a condition to this observation (y/n)? [no]

! Add another observation (y/n)? [no]

! Map a quantity to the surface of the SDF (y/n)? [no]

! Up to which degree should the SDFs be smoothened (0=not at all)? [?]

! Calculate SDF values in nm^-3 (n) or relative to uniform density (y)? [no]

! In which trajectory frame to start processing the trajectory? [1]

! How many trajectory frames to read (from this position on)? [all]

! Use every n-th read trajectory frame for the analysis: [1]
```

(d)

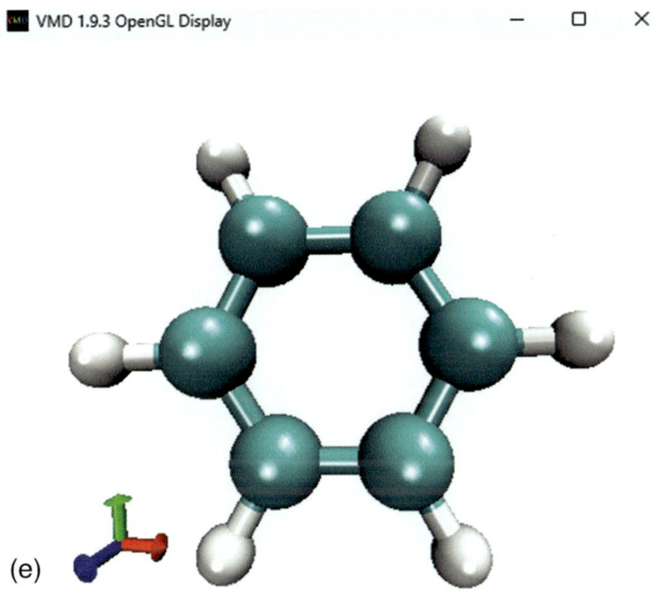

(e)

FIGURE 7.28 (*Continued*) Spatial distribution function (SDF) for benzene-water system in VMD. (a) 'Save coordinates' extension for benzene-water system in VMD. (b) Coordinate file saving for benzene-water system. (c) Trajectory coordinate file loading for benzene-water system in TRAVIS. (d) Input steps for SDF calculation for benzene-water system in TRAVIS. (e) Reference molecule of benzene in the VMD display. (f) Isosurface view of reference molecule of benzene in the VMD display. (g) 'Graphical Representations' window for benzene-water SDF. (h) SDF plot of water around benzene as central molecule in VMD.

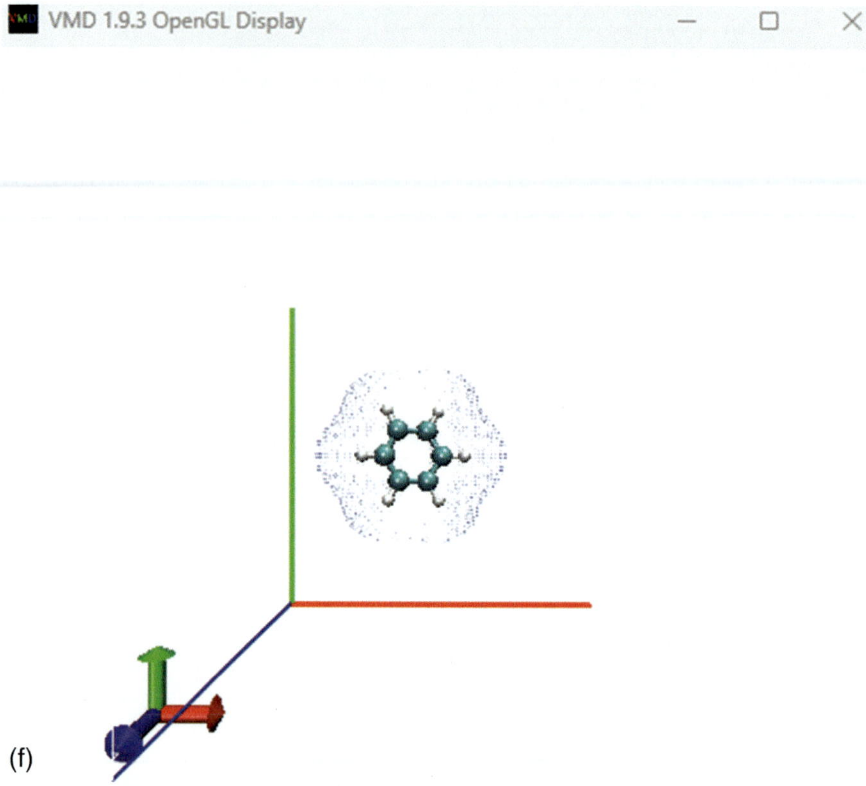

(f)

FIGURE 7.28 (*Continued*) Spatial distribution function (SDF) for benzene-water system in VMD. (a) 'Save coordinates' extension for benzene-water system in VMD. (b) Coordinate file saving for benzene-water system. (c) Trajectory coordinate file loading for benzene-water system in TRAVIS. (d) Input steps for SDF calculation for benzene-water system in TRAVIS. (e) Reference molecule of benzene in the VMD display. (f) Isosurface view of reference molecule of benzene in the VMD display. (g) 'Graphical Representations' window for benzene-water SDF. (h) SDF plot of water around benzene as central molecule in VMD.

(g)

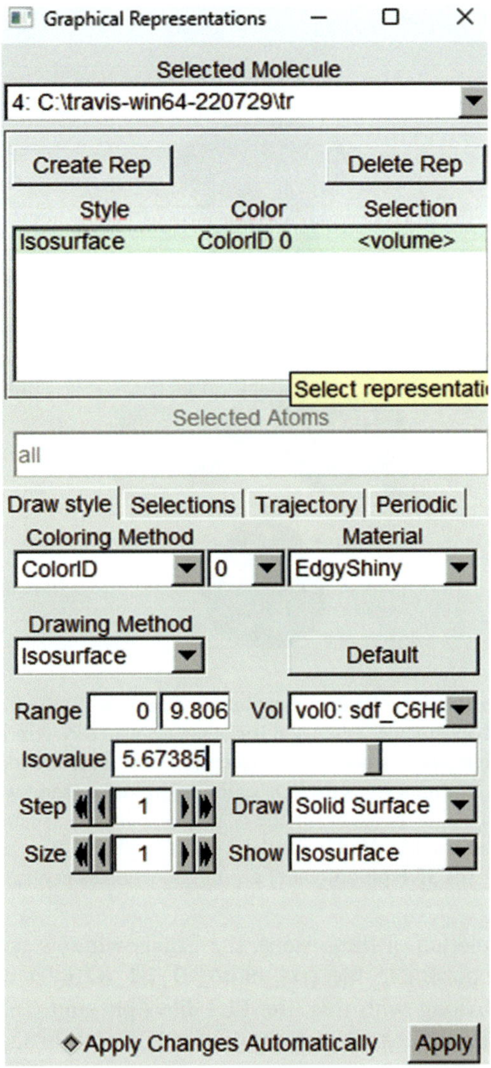

FIGURE 7.28 (*Continued*) Spatial distribution function (SDF) for benzene-water system in VMD. (a) 'Save coordinates' extension for benzene-water system in VMD. (b) Coordinate file saving for benzene-water system. (c) Trajectory coordinate file loading for benzene-water system in TRAVIS. (d) Input steps for SDF calculation for benzene-water system in TRAVIS. (e) Reference molecule of benzene in the VMD display. (f) Isosurface view of reference molecule of benzene in the VMD display. (g) 'Graphical Representations' window for benzene-water SDF. (h) SDF plot of water around benzene as central molecule in VMD.

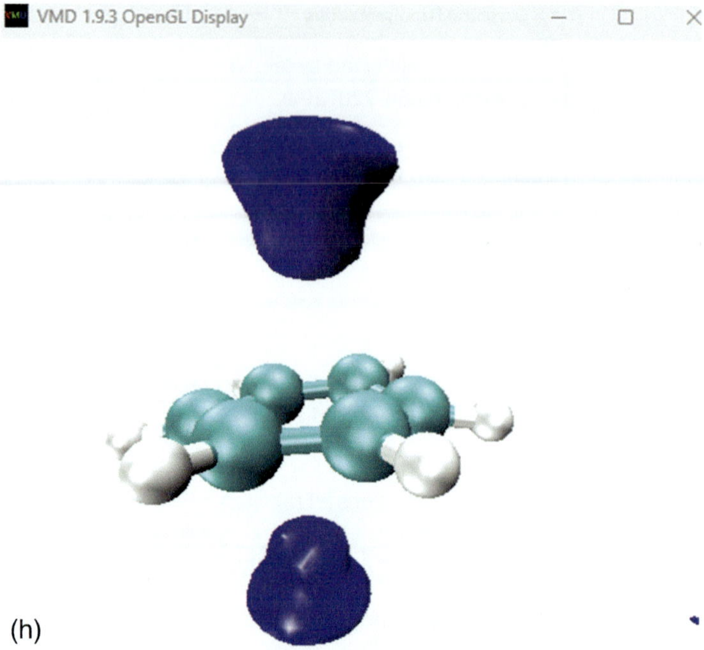

(h)

FIGURE 7.28 (*Continued*) Spatial distribution function (SDF) for benzene-water system in VMD. (a) 'Save coordinates' extension for benzene-water system in VMD. (b) Coordinate file saving for benzene-water system. (c) Trajectory coordinate file loading for benzene-water system in TRAVIS. (d) Input steps for SDF calculation for benzene-water system in TRAVIS. (e) Reference molecule of benzene in the VMD display. (f) Isosurface view of reference molecule of benzene in the VMD display. (g) 'Graphical Representations' window for benzene-water SDF. (h) SDF plot of water around benzene as central molecule in VMD.

After the completion of these steps, the Travis window will automatically close and a reference coordinate file (**ref_c6h6_c1_h1_h2.xyz**) will be created in the Travis directory. Along with this, the PLT file (.plt) and Cube file (.cube) will be generated. Using these files, we can obtain the SDF plot through VMD. Here are the steps involved in this.

Open VMD → Drag and drop the coordinate file (**ref_c6h6_c1_h1_h2.xyz**) to VMD Main window → Open **Graphical Representations** window from **Graphics** tab → Select **CPK** as the **Drawing Method** → The reference molecule of benzene will appear in the VMD display (Figure 7.28e).

To continue the process, drag and drop the coordinate file Open VMD → Drag and drop the PLT file (.plt) to **VMD Main** window → Initially, it appears as Figure 7.28f → Open the **Graphical Representations** window → Select **ColorID 0** to represent the color "**blue**" for water isosurface → Adjust the **Isovalue** using the adjustment bar for proper visualization → Select **Solid Surface** and **Isosurface** to specify **Draw** and **Show** options, respectively.

The graphical representations window is presented in Figure 7.28g, and the final SDF plot is displayed in Figure 7.28h.

The blue iso-surfaces surrounding the benzene molecule indicate the probable locations of the dodecane molecules. These regions suggest areas with a high likelihood of dodecane molecules. Surfaces generated with higher iso values generally indicate areas of greater density or probability. This suggests that the property being measured, such as the presence of a solvent, is more concentrated or more likely to be found in these regions. The distribution is not uniform, indicating preferential solvation sites around the acetone molecule.

7.3.2 Nonbonded Interaction Energy

For nonbonded interaction energy calculation in VMD, the following steps are followed:

Click on **Extensions** → Go to **Analysis** → Select **NAMD Energy** from the drop-down list → The **NAMD Energy** window will appear.

In **NAMD Energy** window, select **.prmtop** file in Molecule, write **resname BEN** (for benzene) as **Selection 1** and **resname wtr** (for water) as **Selection 2.** Tick the boxes of **VDW, Elec, and Nonbond** that respectively indicate van der Waals, electrostatic, and total nonbonded interactions. Set **Output File** name (e.g. BEN-WTR-IE). Select the parameter file and XSC file by browsing. All the other values will be kept as default. Then click on **Run NAMD Energy.** This will run the NAMD energy plugin and after a while the following output (Figure 7.29b) file will be generated:

(a)

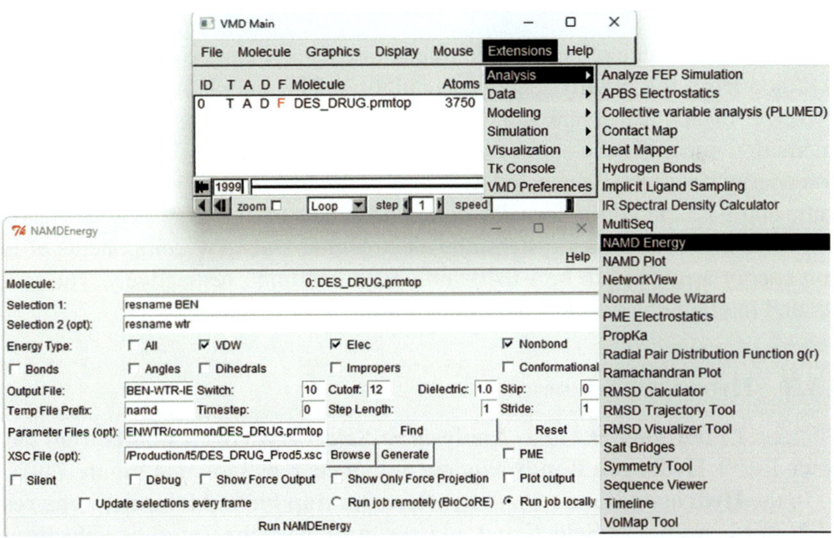

FIGURE 7.29 Non-bonded interaction energy (NAMD energy) protocol for benzene-water system in VMD. (a) NAMD energy inputs for benzene-water system. (b) NAMD energy output for benzene-water system.

(Continued)

BEN-WTR-IE

Frame	Time	Elec	VdW	Nonbond	Total
0	0	-185.995	-188.723	-374.718	-374.718
1	1	-195.135	-190.353	-385.488	-385.488
2	2	-199.765	-188.863	-388.628	-388.628
3	3	-189.05	-193.808	-382.858	-382.858
4	4	-210.894	-197.132	-408.026	-408.026
5	5	-184.901	-187.863	-372.765	-372.765
6	6	-200.31	-189.601	-389.912	-389.912
7	7	-168.552	-211.78	-380.332	-380.332
8	8	-229.596	-198.878	-428.475	-428.475
9	9	-189.264	-207.229	-396.493	-396.493
10	10	-182.085	-190.37	-372.454	-372.454
11	11	-187.857	-182.851	-370.708	-370.708
12	12	-187.076	-193.744	-380.819	-380.819
13	13	-194.95	-176.222	-371.172	-371.172
14	14	-182.184	-195.289	-377.473	-377.473
15	15	-188.501	-191.116	-379.617	-379.617
16	16	-170.957	-187.689	-358.646	-358.646
17	17	-198.902	-192.705	-391.607	-391.607
18	18	-185.118	-195.379	-380.497	-380.497
19	19	-180.567	-180.768	-361.336	-361.336
20	20	-221.626	-199.126	-420.753	-420.753
21	21	-209.038	-189.256	-398.294	-398.294
22	22	-194.425	-189.028	-383.453	-383.453
23	23	-208.854	-198.53	-407.384	-407.384
24	24	-224.259	-208.399	-432.658	-432.658
25	25	-175.708	-188.877	-364.585	-364.585
26	26	-207.946	-174.461	-382.406	-382.406
27	27	-198.164	-194.112	-392.276	-392.276
28	28	-187.318	-188.687	-376.005	-376.005
29	29	-168.744	-184.903	-353.647	-353.647
30	30	-186.348	-196.435	-382.783	-382.783

(b)

FIGURE 7.29 *(Continued)* Non-bonded interaction energy (NAMD energy) protocol for benzene-water system in VMD. (a) NAMD energy inputs for benzene-water system. (b) NAMD energy output for benzene-water system.

The average value of nonbonded interaction energy can be obtained from the data obtained from the NAMD energy calculation. The interaction energy is classified into electrostatic (E_{elec}) and van der Waals (E_{vdW}) components. The total nonbonded interaction energy is the sum of both of these components. The nonbonded interaction energy between acetone and methanol is found to be –2.40 kcal/mol for electrostatic and –1.52 kcal/mol for vdW component. The total energy is –3.92 kcal/mol.

For the benzene-water system, the electrostatic and vdW components of interaction energy are found to be –0.40 and –0.39 kcal/mol, respectively. The total nonbonded interaction energy is –0.79 kcal/mol.

7.3.3 Hydrogen Bonding

Click on **Extensions** → Go to **Analysis** → Select **Hydrogen Bonds** from the drop-down list → **Hydrogen Bonds** window will appear, as shown in Figure 7.30a.

In the **Hydrogen Bonds** window, select **.prmtop** file in Molecule, write **resname BEN** (for benzene) as **Selection 1** and **resname wtr** (for water) as **Selection 2.** Set **Donor-Acceptor distance** (Å) value as 3.0, and **Angle cutoff (degrees)** value as 180. These are the general criteria for hydrogen bonding. Tick the boxes of **Plot the data with MultiPlot,** and choose the output directory by browsing. Specify the output filename. All the other values will be kept as default. Then click on **Find hydrogen**

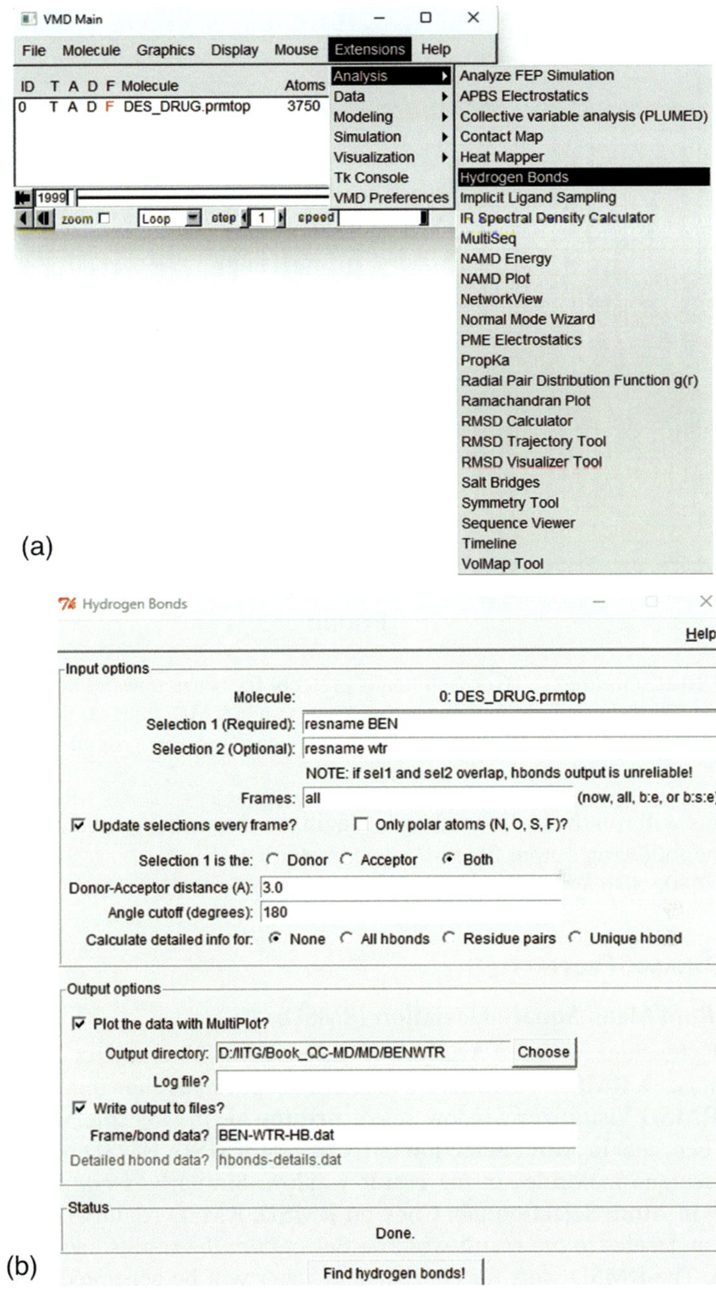

FIGURE 7.30 Hydrogen bonds protocol for benzene-water system in VMD. (a) 'Hydrogen Bonds' extension for benzene-water system in VMD. (b) Hydrogen bond inputs for benzene and water. (c) Hydrogen bonds output for benzene-water system.

(Continued)

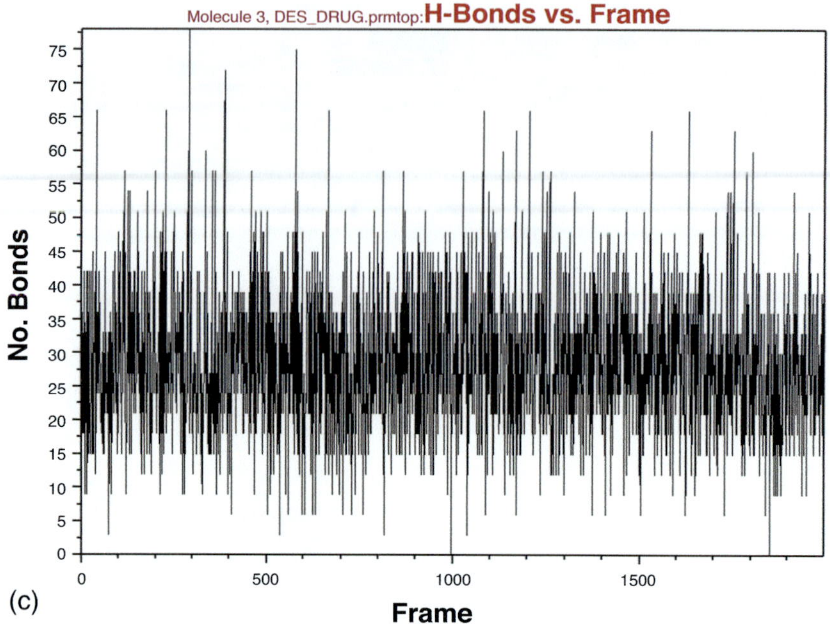

(c)

FIGURE 7.30 (*Continued*) Hydrogen bonds protocol for benzene-water system in VMD. (a) 'Hydrogen Bonds' extension for benzene-water system in VMD. (b) Hydrogen bond inputs for benzene and water. (c) Hydrogen bonds output for benzene-water system.

bonds. This will run the hydrogen bonds plugin, as shown in Figure 7.30b, and after a while, the following output file will be generated:

Figure 7.30c displays the number of hydrogen bonds vs the frame number plot.

7.3.4 Dynamic Property

7.3.4.1 Root Mean Square Deviation (RMSD)

Click on **Extensions** → Go to **Analysis** → Select **RMSD Visualizer Tool** from the dropdown list → **RMSD Visualizer** window will appear, as shown in Figure 7.31b.

In the **RMSD Visualizer** window, select **.prmtop** file in Molecule, write **resname BEN** (for benzene) in **Atom Selection** tab. Click on **RMSD.** RMSD result will generate for acetone molecules in the **result** window. Similarly, **write resname wtr** (for water) in **Atom Selection** tab. Click on **RMSD.** RMSD result will generate for methanol molecules in the **result** window. Select both the results and click on **Plot result** tab. The RMSD plots for benzene and water will be generated, as shown in Figure 7.31c.

The "BEN" component, shown by the blue curve, shows a rapid escalation in RMSD values, ultimately stabilizing at about 25 Å throughout the early phases of the simulation. According to the results, "BEN" undergoes a rapid structural change initially to attain a stable condition. The root mean square deviation (RMSD) is

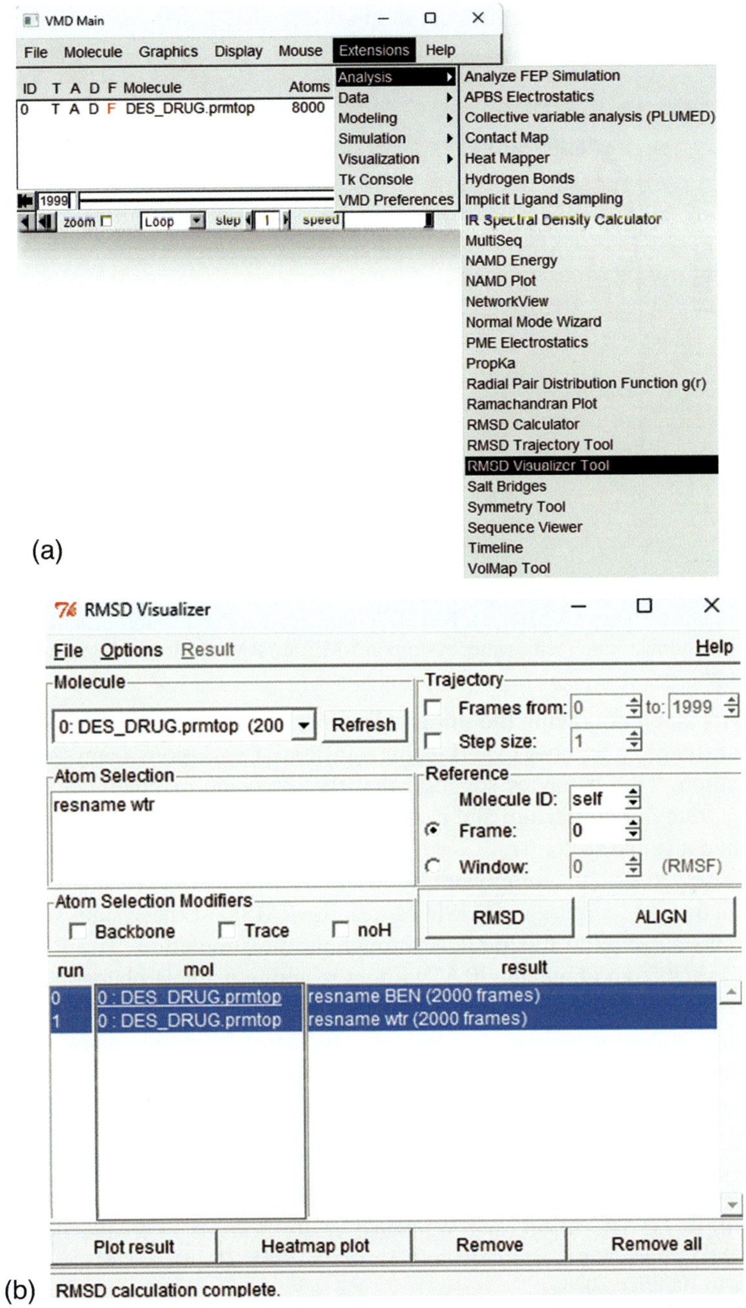

FIGURE 7.31 Root Mean Square Deviation (RMSD) protocol for benzene-water system in VMD. (a) 'RMSD Visualizer Tool' extension in VMD. (b) RMSD visualizer window for benzene-water system in VMD. (c) RMSD plot for benzene (blue)-water (red) system.

(Continued)

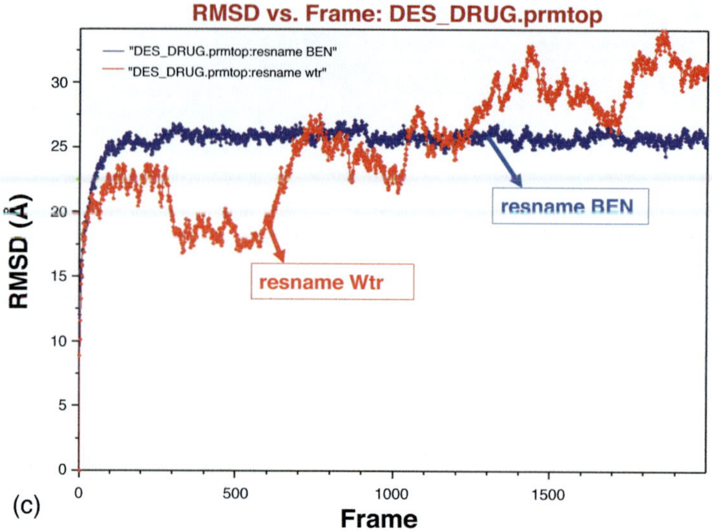

FIGURE 7.31 (*Continued*) Root Mean Square Deviation (RMSD) protocol for benzene-water system in VMD. (a) 'RMSD Visualizer Tool' extension in VMD. (b) RMSD visualizer window for benzene-water system in VMD. (c) RMSD plot for benzene (blue)-water (red) system.

continually stable following the initial adjustment, indicating that "BEN" retains its structural integrity without experiencing substantial variations from this stabilized conformation. This behavior suggests that the molecule or chemical complex has reached a state of equilibrium and is experiencing minimal changes in its conformation around a stable structure.

Conversely, the "wtr" component, shown in red, has distinct behavior. Although it initially shows an increase in RMSD values, the RMSD of the system stays elevated and exhibits substantial fluctuations throughout the simulation. The RMSD values for "wtr" vary from about 18–30 Å, without reaching a stable plateau. The ongoing variations indicate that "wtr" has more structural adaptability, perhaps exploring a wider range of conformations. The lack of a distinct stabilization phase suggests that "wtr" may fundamentally exhibit more dynamism as a result of its interactions with the surrounding environment or intrinsic characteristics.

REFERENCES

1. Phillips, J.C., R. Braun, and W. Wang, et al., Scalable molecular dynamics with NAMD. *Journal of Computational Chemistry*, **2005**. **26**(16): p. 1781–1802. https://doi.org/10.1002/jcc.20289.
2. Frisch, M.J., G.W. Trucks, and H.B. Schlegel, et al., *Gaussian 16*. Gaussian, Inc.: Wallingford CT, 2016. https://gaussian.com/citation/.
3. Bayly, C.I., P. Cieplak, and W. Cornell, et al., A well-behaved electrostatic potential based method using charge restraints for deriving atomic charges: The RESP model. *The Journal of Physical Chemistry*, **1993**. **97**(40): p. 10269–10280. https://doi.org/10.1021/j100142a004.

4. Case, D.A., K. Belfon, and I.Y. Ben-Shalom, et al., *AMBER 14*. University of California: San Francisco, 2014.
5. Wang, J., W. Wang, and P.A. Kollman, et al., Automatic atom type and bond type perception in molecular mechanical calculations. *Journal of Molecular Graphics Model*, **2006**. **25**(2): p. 247–260. https://doi.org/10.1016/j.jmgm.2005.12.005.
6. Martinez, L., R. Andrade, and E.G. Birgin, et al., PACKMOL: A package for building initial configurations for molecular dynamics simulations. *Journal of Computational Chemistry*, **2009**. **30**(13): p. 2157–2164. https://doi.org/10.1002/jcc.21224.
7. Humphrey, W., A. Dalke, and K. Schulten, VMD: Visual molecular dynamics. *Journal of Molecular Graphics*, **1996**. **14**(1): p. 33–38. https://doi.org/10.1016/0263-7855(96)00018-5.
8. Brehm, M. and B. Kirchner, TRAVIS – A free analyzer and visualizer for Monte Carlo and molecular dynamics trajectories. *Journal of Chemical Information and Modeling*, 2011, **51**(8): p. 2007–2023. https://doi.org/10.1021/ci200217w.

Index

Note: **Bold** page numbers refer to tables and *italic* page numbers refer to figures.